Synthesis Lectures on Engineering, Science, and Technology

The focus of this series is general topics, and applications about, and for, engineers and scientists on a wide array of applications, methods and advances. Most titles cover subjects such as professional development, education, and study skills, as well as basic introductory undergraduate material and other topics appropriate for a broader and less technical audience.

Thomas Grurl · Jürgen Fuß · Robert Wille

Noise-Aware Quantum Circuit Simulation with Decision Diagrams

Springer

Thomas Grurl
University of Applied Sciences Upper Austria
Wels, Austria

Jürgen Fuß
University of Applied Sciences Upper Austria
Wels, Austria

Robert Wille
Software Competence Center Hagenberg
Hagenberg, Austria

Technical University of Munich
Munich, Germany

ISSN 2690-0300   ISSN 2690-0327 (electronic)
Synthesis Lectures on Engineering, Science, and Technology
ISBN 978-3-031-71035-3   ISBN 978-3-031-71036-0 (eBook)
https://doi.org/10.1007/978-3-031-71036-0

This Springer imprint is published by the registered company Springer Nature Switzerland AG
The registered company address is: Gewerbestrasse 11, 6330 Cham, Switzerland

If disposing of this product, please recycle the paper.

Preface

In the early '80s, Richard Feynman proposed using a quantum computer in order to simulate quantum systems. By this, he was the first to suggest using a quantum computer to efficiently solve a task, that is exponentially hard for classical computers. Since then, a lot of research has been conduced on the development of quantum hardware and new quantum algorithms. Today, many more applications for quantum computers have been identified, and first quantum computers are already commercially available. With the increasing interested in quantum computing, the need for quantum circuit simulators, i.e., tools that simulate the execution of a quantum circuit on classical hardware, has also risen, as these simulators are an important tool for the development of new quantum algorithms. However, the task of quantum circuit simulation is exponentially hard on classical machines. Worse still, today's quantum computers are prone to noise effects (i.e., errors), that obscure their calculations. Taking these noise effects into account during quantum circuit simulation makes the problem even more complex, but is essential to understand how algorithms behave when executed on real hardware.

This book proposes to tackle the complexity of *noise-aware* quantum circuit simulation using *decision diagrams*. Decision diagrams are a data structure for efficiently representing functionality, that has already been successfully used for other design tasks, yet, their potential for noise-aware quantum circuit simulation has been largely unexplored. This is changed in this book.

To this end, we review the current state-of-the-art of decision diagram-based quantum circuit simulation and propose dedicated extensions for handling errors. Next, we present and implement two complementary decision diagram-based noise-aware quantum circuit simulation approaches. Subsequent evaluations against industry-grade simulators demonstrate the viability of the proposed solutions with substantial speed-ups for many applications. Finally, we demonstrate the necessity of noise-aware quantum circuit simulation with on one use-case: quantum error correction. A framework is proposed that automatically applies quantum error correction codes to quantum circuits, followed by a noise-aware quantum circuit simulation to evaluate the reliability of the error-correcting code. Overall, this book demonstrates the necessity for noise-aware quantum circuit simulation and confirms the utility of decision diagrams for this task. All code developed within the scope of this book is available as open-source.

This book is the result of several years of research arising from a collaboration between the Johannes Kepler University Linz, the University of Applied Sciences Upper Austria, and the Technical University of Munich. We express our gratitude to all institutions for enabling this productive partnership and for fostering an intellectually stimulating and enjoyable environment where scientific ideas can thrive. Special thanks go to our colleagues, including Stefan Hillmich, Lukas Burgholzer, and Richard Küng, with whom we engaged in many insightful discussions. Furthermore, this endeavour would not have been possible without the support of the State of Upper Austria, which funded this research with the University of Applied Sciences Ph.D. program (managed by the FFG), the European Research Council (ERC) under the European Union's Horizon 2020 research and innovation program (grant agreement No. 101001318), and the Munich Quantum Valley, which is supported by the Bavarian state government with funds from the Hightech Agenda Bayern Plus. Finally, we would like to thank Springer Nature and especially Charles "Chuck" Glaser for publishing this work.

Hagenberg, Austria   Thomas Grurl
Munich, Germany   Jürgen Fuß
July 2024   Robert Wille

Contents

List of Figures

List of Tables

Part I
Introduction and Background

Introduction

The discovery of quantum mechanics in the early 20th century by physicists such as Max Planck, Albert Einstein, Niels Bohr, Werner Heisenberg, and Erwin Schrödinger revolutionized physics. In turn, quantum computers, i.e., computers that exploit quantum mechanical effects, are expected to revolutionize computing by solving problems that are intractable for today's computers. The idea of exploiting quantum physics for computing dates back to 1980 when Paul Benioff suggested building a computer based on quantum mechanical principles [10]. Shortly thereafter, Richard Feynman proposed using such a quantum computer to efficiently simulate quantum systems, a task that is exponentially hard for classical computers [37]. One decade later, in 1992, David Deutsch and Richard Jozsa presented one of the first quantum algorithms that could solve a problem exponentially faster than a classical computer [29]. However, since the Deutsch-Jozsa algorithm solves an artificial problem only, it did not raise much interest in actually building a quantum computer. This drastically changed in 1994, when Peter Shor proposed an algorithm to solve a very relevant problem exponentially faster compared to the best known classical algorithm, namely, factoring numbers [95]. Only one year later, he also presented the first quantum error correction algorithm [97].

Since then, many more quantum algorithms have been found in areas such as machine learning, chemistry, or quantum optimization [12, 22, 33, 60, 91]. Among the many applications of these quantum algorithms is, for example, making the chemical processes of creating ammonia more efficient, i.e. the Haber-Bosch reaction, which currently consumes 1–2% of the world's total energy [23].

Quantum computers achieve these impressive speedups by exploiting quantum mechanical effects in their computations. More precisely, in contrast to classical computers which use bits that can assume the states 0 or 1, a quantum computer works on quantum bits

T. Grurl et al., *Noise-Aware Quantum Circuit Simulation with Decision Diagrams*, Synthesis Lectures on Engineering, Science, and Technology, https://doi.org/10.1007/978-3-031-71036-0_1

(*qubits* for short). Qubits can also assume the states 0 and 1, which are called basis states and—using Dirac notation—are written as $|0\rangle$ and $|1\rangle$, but additionally they can be in an arbitrary linear combination of these two basis states. This effect is called *superposition* and together with the effect that qubits can be *entangled*, a quantum register of n qubits can be in an arbitrary combination of up to 2^n states at the same time. This is in contrast to classical registers, which can only assume one state at a time. While this vast complexity brings with it a huge potential for solving computational problems, it also leads to tremendous challenges in the development and design of quantum hardware.

However, even despite these challenges, considerable milestones have already been reached in the development of quantum hardware. Among these milestones that have already been reached are IBM's first commercial quantum computer in 2019 [62], Alphabet's claim in the same year to have demonstrated quantum advantage [5][1], IBM's announcement of a quantum chip with 433 qubits in 2022 [61], or Alphabet's announcement in 2023 to have successfully reduced the noisiness of a quantum system through error correction, thereby achieving a huge milestone in building large scale and error-prone quantum hardware [3]. All this work is driven by big players (such as Alphabet (Google), AWS, IBM, Intel, Microsoft), start-ups (such as IonQ, Quantinuum, Rigetti, or Zapata), and, finally, many research institutes and universities from all over the world.

Despite these impressive achievements towards the physical realization of quantum hardware, quantum computers are still an emerging technology with limited reliability and accessibility. Therefore, a substantial amount of research on quantum computing still depends on so-called quantum circuit simulators. As their name suggests, they simulate the execution of a quantum circuit on classical hardware and, by this, allow for the development and evaluation of quantum circuits without access to actual quantum hardware. Moreover, quantum circuit simulators allow for more insight into the considered quantum applications, since their results not only provide probabilistic measurements (like a real quantum computer does) but also offer full access to the quantum state.

The task of quantum circuit simulation is conceptually simple on classical hardware and boils down to matrix-vector-multiplication, with vectors representing quantum states and matrices representing quantum operations. However, the respective vectors are of size 2^n with n being the number of tracked qubits, which makes the simulation of quantum registers consisting of more than a few qubits extremely challenging—the current practical limit being around 50 qubits [27]. This might also be one of the reasons why many quantum circuit simulators (such as [79, 85, 87, 103, 108, 121]) mimic only *perfect* quantum computers.

Yet, due to the inherent fragility of quantum mechanical effects, quantum computers are prone to noise effects (i.e., errors) that obscure their calculations [88]. While error mitigation is constantly improving, errors generated by noise effects are still an overshadowing aspect

---

[1] This claim has since been made by many more entities.

of quantum computing. Hence, considering those noise effects during quantum circuit simulation is essential to gain insight into the behavior of algorithms when they are executed on real quantum hardware.

Fortunately, noise effects in quantum computers have been well studied and mathematical models to approximate them are available [82]. However, employing these models during quantum circuit simulation makes the hard problem of quantum circuit simulation even harder, limiting corresponding approaches (see, e.g., [7, 24, 25, 41, 54, 64, 66, 73, 89, 98, 109, 111]). More precisely, the main challenge of quantum circuit simulation is already caused by the exponential nature of the vectors and matrices that are required to describe quantum states and operations. Additionally, considering errors caused by noise effects requires extended descriptions and/or methods (covered in more detail later in this book) that add substantial further complexity. How to efficiently cope with all this complexity is an important research problem.

An interesting approach to tackle the complexity of the quantum world is based on the use of *decision diagrams* for quantum computing (as introduced, e.g., in [121]). While originally developed to solve problems in the classical realm of design automation [9, 18, 19, 31, 40, 81, 112], they have already proven their usability to represent quantum objects, as they offer a compact data structure that is often below the exponential size of other solutions. They have been successfully used for quantum computing design tasks such as circuit verification (e.g., [20, 84, 99, 100, 110]), gate synthesis (e.g., [34, 83, 99, 123]), and quantum circuit simulation (e.g., [2, 44, 79, 80, 85, 94, 105–107, 121]).

However, to date, their potential for noise-aware quantum circuit simulation has been largely unexplored. Hence, it remains unknown whether the compact representation provided by current state-of-the-art quantum decision diagrams can also be utilized, when noise is considered during the simulation. This question is addressed in this book.

To this end, the next part sets the *foundations* of this investigation on noise-aware quantum circuit simulation by reviewing the current state-of-the-art on quantum decision diagrams and proposing dedicated decision diagram structures to represent noisy quantum states:

- Chapter 3 (based on [47, 49]) reviews and evaluates the current state-of-the-art of quantum decision diagrams. More precisely, first, the cost of multiplication and addition of decision diagrams—essential operations for quantum circuit simulation—is evaluated. Although the cost of both operations strongly correlates with the number of nodes of the involved decision diagrams, the actual cost of multiplication can still vary considerably depending on the structure of the decision diagrams. In contrast, the cost of adding decision diagrams depends only on the number of nodes. Next, the performance of a state-of-the-art decision diagram-based quantum circuit simulator is compared with the industry-grade matrix-vector-multiplication (i.e., array-based) simulator from Atos. The results show that the decision diagram-based solution performs magnitudes faster, while using less hardware resources, for many quantum algorithms (e.g., for Shor's algorithm to factor integers). By this, the chapter confirms the utility of decision diagrams for quantum computing and serves as starting point for the remainder of this book.

- Chapter 4 (based on [51]) explicitly presents an optimized decision diagram representation for density matrices. Density matrices incorporate all possible states into a single description and are therefore an indispensable tool for accurately describing noisy quantum states. In an empirical evaluation, the proposed density matrix representation is then compared with the established decision diagram matrix representation, showing that the optimized representation is never larger and up to 50% more compact in the best case. The smaller size of the proposed decision diagram representation results in runtime improvements for quantum circuit simulation of up to 53%. This more efficient representation for noisy quantum states serves as the basis for the noise-aware quantum circuit simulator developed in the next chapter.

Based on the earlier chapters, two complementary *approaches* for noise-aware quantum circuit simulation are developed that utilize decision diagrams, which are presented next:

- Chapter 5 (based on [48, 50]) investigates the potential of decision diagrams for *deterministic* noise-aware quantum circuit simulation using density matrices and Kraus operators. An advanced scheme is proposed to apply errors compared to the straightforward implementation of the formalism. The proposed simulator is implemented and evaluated in a comprehensive manner against industry-grade simulators from Atos and Qiskit. The results are mixed, showing that the proposed solution is magnitudes faster for one considered quantum circuit, while at the same time, being slower for most of the others.
- Chapter 6 (based on [50, 52]) investigates a complementary approach for noise-aware quantum circuit simulation by considering errors in a *stochastic* fashion, i.e., errors are assumed to occur randomly throughout individual simulation runs. Subsequently, the true effects of errors are approximated, forming empirical averages over multiple simulation runs (Monte Carlo). This provides a conceptually suitable and mathematically rigorous solution for the classical simulation of noisy quantum computations. The proposed solution is implemented on top of a decision diagram-based simulator in an advanced manner and evaluated against industry-grade simulators from Atos and Qiskit. The results show performance improvements for many of the considered quantum circuits, with runtime improvements in the order of magnitudes.

Having noise-aware quantum circuit simulators allows one to test quantum circuits under realistic conditions. This is essential for some *applications* as shown in the next part:

- Chapter 7 (based on [53]) offers an important application for the decision diagram-based noise-aware quantum circuit simulators proposed in the previous chapters, namely, a framework for evaluating quantum error-correcting codes. Error correction is a necessary prerequisite for executing large and complex quantum circuits. Consequently, a lot of research is being carried out to further develop such schemes. However, much of this research relies heavily on manual labor and/or is based on theoretical results only. The chapter proposes an open-source framework that supports engineers and researchers in the task of evaluating error-correcting codes. The framework allows one to automatically apply error correction schemes to a given quantum circuit, followed by an automatic noise-aware quantum circuit simulation to evaluate its reliability. Case studies show that the proposed framework allows efficient evaluations of error-correcting codes and demonstrate the necessity for noise-aware quantum circuit simulation.

To keep this book self-contained, Chap. 2 reviews the concepts of quantum computing and noise in quantum computing. Chapter 8 concludes this book.

During the pursuit of these ideas, the corresponding contributions have been published at international conferences (i.e., ICCAD [51], ISMVL [47, 49, 51], DAC [113], VLSID [53], SPIN [72], and DATE [52]), as well as in a leading journal (i.e., TCAD [50]). All implementations that were developed in the context of this book have been published as part of the *Munich Quantum Toolkit* (MQT [115]; formerly known as JKQ [114]) and the source code of all developed tools is available under MIT license under https://github.com/cda-tum/.

Background

2

In order to keep this book self-contained, this section reviews the basic concepts of quantum computing as well as noise effects occurring in today's quantum hardware.

2.1 Quantum Computing

2.1.1 Quantum States

In the classical world, the basic unit of information is a bit, which can either assume the state 0 or 1. In the quantum world, the smallest unit of information is called a *quantum bit* or *qubit*. Similar to a classical bit, a qubit can assume the states 0 and 1, referred to as *basis states* and—using Dirac notation—are written as $|0\rangle$ and $|1\rangle$. Additionally, a qubit can also assume an almost arbitrary combination of the two basis states, which is then called a *superposition*. More precisely, the state of the qubit $|\psi\rangle$ is written as

$$|\psi\rangle = \alpha_0 \cdot |0\rangle + \alpha_1 \cdot |1\rangle \text{ with } \alpha_0, \alpha_1 \in \mathbb{C} \text{ where } |\alpha_0|^2 + |\alpha_1|^2 = 1. \qquad (2.1)$$

The values α_0 and α_1 are called *amplitudes* and describe how strongly the qubit is related to each of the basis states. The state of a single qubit can be visualized as a vector on the Bloch sphere as shown in Fig. 2.1. Here, the state is defined by the angles λ and θ

$$|\psi\rangle = \cos\left(\frac{\theta}{2}\right) \cdot |0\rangle + e^{i\lambda} \sin\left(\frac{\theta}{2}\right) \cdot |1\rangle. \qquad (2.2)$$

© The Author(s), under exclusive license to Springer Nature Switzerland AG 2025
T. Grurl et al., *Noise-Aware Quantum Circuit Simulation with Decision Diagrams*,
Synthesis Lectures on Engineering, Science, and Technology,
https://doi.org/10.1007/978-3-031-71036-0_2

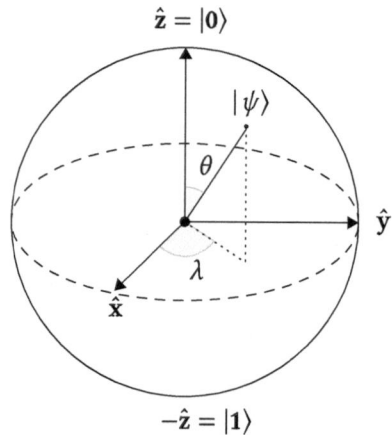

Fig. 2.1 Bloch sphere

In contrast to the classical world, where bits may be observed without changing them, observing a qubit, i.e., measuring it, collapses the qubit to a basis state. More precisely, measuring a qubit collapses it to $|0\rangle$ with probability $|\alpha_0|^2$ and to $|1\rangle$ with probability $|\alpha_1|^2$, respectively. The normalization constraint on the amplitudes ($|\alpha_0|^2 + |\alpha_1|^2 = 1$) ensures that the summed probabilities are 1. After the measurement, the qubit stays in the measured basis state.

Single qubits can be combined into larger quantum states (i.e., quantum systems) that are then called *quantum registers*. The formalism to describe one qubit can be generalized to multi-qubit systems: Each qubit can assume two basis states and a quantum register of n qubits has therefore 2^n possible basis states $|x\rangle \in \{1, 0\}^n$. An amplitude for each basis state describes how much the quantum state relates to this specific basis state.

Definition 2.1 All possible pure states of a quantum system composed of n qubits are defined by

$$|\psi\rangle = \sum_{x \in \{0,1\}^n} \alpha_x \cdot |x\rangle, \text{ where } \sum_{x \in \{0,1\}^n} |\alpha_x|^2 = 1 \text{ and } \alpha_x \in \mathbb{C} \text{ for all } |x\rangle \in \{1, 0\}^n. \quad (2.3)$$

The state $|\psi\rangle$ of a quantum register of size n can also be written as a complex column vector of size 2^n, whose entries correspond to the amplitudes α_x for $x \in \{0, 1\}^n$.

The order of qubits in a quantum register is inherently arbitrary since it can be simply changed by a permutation operation. Without loss of generality, in this book, the first (i.e., left most) qubit is assumed to be the most significant one, analogous to the binary representation of natural numbers. Thus, the qubits of an n-qubit register are correspondingly indexed with $q_{n-1}, \ldots, q_2, q_1, q_0$.

Example 2.1 Consider the two-qubit quantum register

$$|\varphi\rangle = 1 \cdot |00\rangle + 0 \cdot |01\rangle + 0 \cdot |10\rangle + 0 \cdot |11\rangle, \tag{2.4}$$

which is represented by the state vector

$$\begin{bmatrix} 1 & 0 & 0 & 0 \end{bmatrix}^{\top}. \tag{2.5}$$

This is a valid state, since it satisfies the normalization constraint $|1|^2 + |0|^2 + |0|^2 + |0|^2 = 1$. Measuring the system yields $|00\rangle$,[1] with probability 1.

2.1.2 Quantum Operations

Quantum states can be manipulated using quantum operations. With the exception of the measurement operation, all quantum operations are inherently reversible and are represented by matrices.

Definition 2.2 A quantum operation U that acts on a register composed of n qubits is expressed as a $2^n \times 2^n$ unitary matrix, i.e., a complex-valued square matrix whose inverse is given by its conjugate transpose, i.e., $UU^{\dagger} = U^{\dagger}U = I$.

Since these operations are reversible, a single-qubit operation can be visualized as rotations and reflections on the Bloch sphere (see Fig. 2.1). Rotations along the X-, Y-, and Z-axis are defined by the matrices

$$RX(\theta) = \begin{bmatrix} \cos\frac{\theta}{2} & -i\sin\frac{\theta}{2} \\ -i\sin\frac{\theta}{2} & \cos\frac{\theta}{2} \end{bmatrix}, RY(\theta) = \begin{bmatrix} \cos\frac{\theta}{2} & -\sin\frac{\theta}{2} \\ \sin\frac{\theta}{2} & \cos\frac{\theta}{2} \end{bmatrix}, \text{ and} \tag{2.6}$$

$$RZ(\theta) = \begin{bmatrix} e^{-i\frac{\theta}{2}} & 0 \\ 0 & e^{i\frac{\theta}{2}} \end{bmatrix}. \tag{2.7}$$

While the Bloch Sphere and these parameterized rotations are helpful in illustrating quantum states and operations, in practice, often more specific operations are used in quantum circuits. Commonly used one-qubit operations are

---

[1] In this book the results of measurements of quantum states are also written in Dirac notation.

$$H = \frac{1}{\sqrt{2}} \cdot \begin{bmatrix} 1 & 1 \\ 1 & -1 \end{bmatrix}, X = \begin{bmatrix} 0 & 1 \\ 1 & 0 \end{bmatrix}, Z = \begin{bmatrix} 1 & 0 \\ 0 & -1 \end{bmatrix}, \tag{2.8}$$

$$Y = iXZ = \begin{bmatrix} 0 & -i \\ i & 0 \end{bmatrix}, T = \begin{bmatrix} 1 & 0 \\ 0 & e^{i\pi/4} \end{bmatrix}, \text{ and } I = \begin{bmatrix} 1 & 0 \\ 0 & 1 \end{bmatrix}, \tag{2.9}$$

with the H (Hadamard) gate transforming a basis state into a superposition, X being the quantum equivalent of the classical NOT operation, the Z gate flipping the phase of a qubit, the Y gate being a combination of the X and Z gate, and the T gate introducing a phase of $\pi/4$. The identity "operation" I leaves a state unchanged and is relevant in the context of simulating noise.

Additionally, there are also multi-qubit operations. A prominent two-qubit operation is the controlled-NOT (CNOT) operation, which negates the state of a qubit, if the chosen control qubit is $|1\rangle$. A CNOT operation with control on the first and target on the second qubit is represented by the matrix

$$CNOT = \begin{bmatrix} 1 & 0 & 0 & 0 \\ 0 & 1 & 0 & 0 \\ 0 & 0 & 0 & 1 \\ 0 & 0 & 1 & 0 \end{bmatrix}. \tag{2.10}$$

2.1.3 Applying Operations

An operation is applied to a quantum state by matrix-vector-multiplication.

Definition 2.3 A quantum operation U transforms a quantum register $|\psi\rangle$ to $|\psi'\rangle$ as defined in Eq. (2.11).

$$|\psi'\rangle = U \cdot |\psi\rangle \tag{2.11}$$

For matrix-vector-multiplication, the size of both the matrix and the vector have to match. Since quantum states often consist of more than one qubit, this often requires "enlarging" the operation, which is done using the Kronecker product with the I operator.

Definition 2.4 The Kronecker product between an n-qubit operation U and an m-qubit operation V is denoted as $U \otimes V$ and calculated like so

$$U \otimes V = \begin{bmatrix} U_{1,1} \cdot V & U_{1,2} \cdot V & \dots & U_{1,2^n} \cdot V \\ U_{2,1} \cdot V & U_{2,2} \cdot V & \dots & U_{2,2^n} \cdot V \\ \vdots & \vdots & \ddots & \vdots \\ U_{2^n,1} \cdot V & U_{2^n,2} \cdot V & \dots & U_{2^n,2^n} \cdot V \end{bmatrix}. \tag{2.12}$$

The resulting matrix is of size $2^{m+n} \times 2^{m+n}$ and operates on a quantum system composed of $m + n$ qubits.

Example 2.2   Consider again the two-qubit quantum register

$$|\varphi\rangle = \begin{bmatrix} 1 & 0 & 0 & 0 \end{bmatrix}^{\top}. \tag{2.13}$$

from Example 2.1. In order to apply an X operation to the rightmost qubit of $|\varphi\rangle$ (q_0), the operation has to be enlarged first using the Kronecker product

$$\underbrace{\begin{bmatrix} 1 & 0 \\ 0 & 1 \end{bmatrix}}_{I} \otimes \underbrace{\begin{bmatrix} 0 & 1 \\ 1 & 0 \end{bmatrix}}_{X} = \underbrace{\begin{bmatrix} 0 & 1 & 0 & 0 \\ 1 & 0 & 0 & 0 \\ 0 & 0 & 0 & 1 \\ 0 & 0 & 1 & 0 \end{bmatrix}}_{I \otimes X}. \tag{2.14}$$

Applying this operation to $|\varphi\rangle$ results in

$$\underbrace{\begin{bmatrix} 0 & 1 & 0 & 0 \\ 1 & 0 & 0 & 0 \\ 0 & 0 & 0 & 1 \\ 0 & 0 & 1 & 0 \end{bmatrix}}_{I \otimes X} \cdot \underbrace{\begin{bmatrix} 1 \\ 0 \\ 0 \\ 0 \end{bmatrix}}_{|\varphi\rangle} = \underbrace{\begin{bmatrix} 0 \\ 1 \\ 0 \\ 0 \end{bmatrix}}_{|\varphi_1\rangle}. \tag{2.15}$$

The processes of applying a H operation to the left most qubit of $|\varphi_1\rangle$ (q_1), is analogous, i.e., the operation has to be enlarged using the Kronecker product

$$\underbrace{\frac{1}{\sqrt{2}} \cdot \begin{bmatrix} 1 & 1 \\ 1 & -1 \end{bmatrix}}_{H} \otimes \underbrace{\begin{bmatrix} 1 & 0 \\ 0 & 1 \end{bmatrix}}_{I} = \underbrace{\frac{1}{\sqrt{2}} \cdot \begin{bmatrix} 1 & 0 & 1 & 0 \\ 0 & 1 & 0 & 1 \\ 1 & 0 & -1 & 0 \\ 0 & 1 & 0 & -1 \end{bmatrix}}_{H \otimes I}. \tag{2.16}$$

and then the operation is applied to $|\varphi_1\rangle$ like so

$$\underbrace{\frac{1}{\sqrt{2}} \cdot \begin{bmatrix} 1 & 0 & 1 & 0 \\ 0 & 1 & 0 & 1 \\ 1 & 0 & -1 & 0 \\ 0 & 1 & 0 & -1 \end{bmatrix}}_{H \otimes I} \cdot \underbrace{\begin{bmatrix} 0 \\ 1 \\ 0 \\ 0 \end{bmatrix}}_{|\varphi_1\rangle} = \underbrace{\begin{bmatrix} 0 \\ \frac{1}{\sqrt{2}} \\ 0 \\ \frac{1}{\sqrt{2}} \end{bmatrix}}_{|\varphi_2\rangle}. \tag{2.17}$$

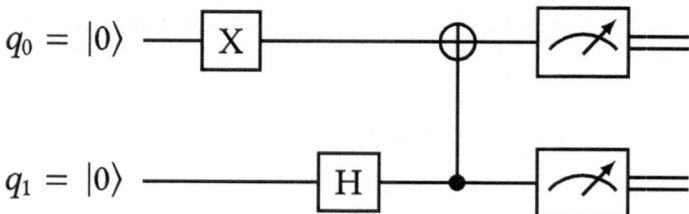

Fig. 2.2 Quantum circuit

Next, a CNOT operation is applied to $|\varphi_2\rangle$, which negates the amplitude of q_0 if q_1 is set to $|1\rangle$. This is given by

$$\underbrace{\begin{bmatrix} 1 & 0 & 0 & 0 \\ 0 & 1 & 0 & 0 \\ 0 & 0 & 0 & 1 \\ 0 & 0 & 1 & 0 \end{bmatrix}}_{\text{CNOT}} \cdot \underbrace{\begin{bmatrix} 0 \\ \frac{1}{\sqrt{2}} \\ 0 \\ \frac{1}{\sqrt{2}} \end{bmatrix}}_{|\varphi_2\rangle} = \underbrace{\begin{bmatrix} 0 \\ \frac{1}{\sqrt{2}} \\ \frac{1}{\sqrt{2}} \\ 0 \end{bmatrix}}_{|\varphi'\rangle} . \tag{2.18}$$

Measuring $|\varphi'\rangle$ either yields $|01\rangle$ or $|10\rangle$, each with probability $|\frac{1}{\sqrt{2}}|^2 = \frac{1}{2}$. Note that the measurement outcome of one qubit affects the other one as well—an essential concept in quantum computing known as *entanglement*.

2.1.4 Quantum Circuit Representation

Quantum circuits, similar to classical circuits, realize some functionality and are often illustrated as circuit diagrams. Diagrams consist of one or more vertical lines that represent qubits. By convention, the top line represents qubit 0 (q_0) and the remaining qubits are labeled sequentially. The state of a quantum circuit with n qubits is represented by the quantum state $|q_{n-1} \ldots q_1 q_0\rangle$. The lines are interrupted by operations that are applied to the respective qubit, which is usually indicated by labeled rectangles. In a quantum circuit, time flows from left to right, i.e., the leftmost operation on a qubit wire is applied first, the second leftmost operation is applied, and so on.

Example 2.3 In Fig. 2.2, a quantum circuit consisting of 2 qubits and 3 quantum operations (without counting the measurements at the end of the circuit) is depicted. The circuit consists of the same operations that are applied in Example 2.2, i.e., to the input state $|00\rangle$ an X operation is applied to q_0, then an H operation is applied to q_1, followed by a CNOT operation with q_1 as the control qubit (denoted by the symbol $\bullet$) and q_0 as the target qubit (denoted by the symbol $\oplus$). Subsequently, the qubits are in state

$$|\varphi'\rangle = \begin{bmatrix} 0 & \frac{1}{\sqrt{2}} & \frac{1}{\sqrt{2}} & 0 \end{bmatrix}^{\top}. \tag{2.19}$$

The final operations are measurements that cause the system to collapse, thus, the final state is $|01\rangle$ or $|10\rangle$, each with probability $|\frac{1}{\sqrt{2}}|^2 = \frac{1}{2}$.

2.2   Noise in Quantum Computing

The formalism presented above can be used to describe how perfect quantum computers behave. However, due to the fragility of quantum mechanical effects, real quantum computers are prone to noise, which causes errors and uncertainty during computations. This section first introduces how uncertainty in quantum states can be represented and how "correctness" of quantum states can be quantified. Afterwards, sources and types of errors are introduced.

2.2.1   Uncertainty in Quantum States

Noise effects can be viewed as (unwanted) operations on the state. However, while ideal quantum operations are deterministic, errors can have an additional degree of randomness. For instance, whenever a specific qubit is used, an error may occur with probability p while no error occurs with probability $1 - p$. The outcome of an erroneous quantum computation cannot be described anymore by a single state vector (i.e., pure state). Instead, it is described by a *mixture* (or ensemble) of possible pure states, i.e.,

Definition 2.5 Given a quantum system that is not completely known but in a mixture of finitely many possible states. This state mixture is fully described by the ensemble of pure states $\{(p_i, |\psi_i\rangle)|1 \leq i \leq k\}$, with the system being in state $|\psi_i\rangle$ with probability p_i and

$$\sum_{i=1}^{k} p_i = 1. \tag{2.20}$$

An important concept when working with noisy quantum states is the *fidelity*, as it allows one to express the correctness of a quantum state as a single number. More precisely, the fidelity is a measure that expresses the overlap between two states [65]. Two identical quantum states (i.e., two completely overlapping states) have a fidelity of 1, while two entirely different quantum states (such as $|0\rangle$ and $|1\rangle$) have a fidelity of 0.

Definition 2.6 The fidelity between the two pure quantum states[2] $|\psi\rangle$ and $|\phi\rangle$ is given by

$$|\langle\psi|\phi\rangle|^2 := |\langle\psi| \cdot |\phi\rangle|^2, \text{ where } \langle\psi| := |\psi\rangle^\dagger \qquad (2.21)$$

Example 2.4 Consider the quantum states $|\varphi_2\rangle$ and $|\varphi'\rangle$ of Example 2.2. The fidelity between the two states is given by

$$\left|\langle\varphi_2|\varphi'\rangle\right|^2 = \left\|\begin{bmatrix} 0 & \frac{1}{\sqrt{2}} & 0 & \frac{1}{\sqrt{2}} \end{bmatrix} \cdot \begin{bmatrix} 0 \\ \frac{1}{\sqrt{2}} \\ \frac{1}{\sqrt{2}} \\ 0 \end{bmatrix}\right\|^2 = \left|\frac{1}{2}\right|^2 = \frac{1}{4}. \qquad (2.22)$$

2.2.2 Noise Effects in Quantum Computers

Noise effects that occur during quantum computing can be classified into two categories [104]:

- *Gate errors* (also known as operational errors)
- *Decoherence errors* (also known as retention errors).

Gate errors are introduced when operations are executed [104]. They occur because quantum computers are mechanical apparatuses that do not always apply operations perfectly. Instead, an operation may not be executed at all or in a (slightly) modified fashion. Since gate errors are highly specific for each quantum computer and even vary for qubits within the quantum computer, they are often approximated by using depolarization errors [66, 89]. A depolarization error describes that a qubit is set to a completely random state [82]. For the publicly available quantum computer ibm_sherbrooke, which is one of IBM's Eagle r3 (Version 1.2.1) processors, the gate error probabilities are currently on the order of 10^{-2} to 10^{-4} [63, 102].

Example 2.5 Consider again the two-qubit quantum register

$$|\varphi'\rangle = \begin{bmatrix} 0 & \frac{1}{\sqrt{2}} & \frac{1}{\sqrt{2}} & 0 \end{bmatrix}^\top = \frac{1}{\sqrt{2}} \cdot (|01\rangle + |10\rangle) \qquad (2.23)$$

from Example 2.2. Suppose that this register might be affected by a gate error in qubit q_0, depolarizing it. With probability $1 - p$, nothing happens and the quantum state remains unchanged, while the qubit becomes depolarized with probability p. This effect can be

---

[2] The fidelity can also be calculated between two mixed states, see, e.g., [82].

captured by applying either I, X, Y, or Z to q_0—each with probability $\frac{p}{4}$ [82]. This produces the mixture

$$\{(1 - p, \tfrac{1}{\sqrt{2}} \cdot (|01\rangle + |10\rangle)), \tag{2.24}$$

$$(\tfrac{p}{4}, \tfrac{1}{\sqrt{2}} \cdot (|01\rangle + |10\rangle)), \tag{2.25}$$

$$(\tfrac{p}{4}, \tfrac{1}{\sqrt{2}} \cdot (|00\rangle + |11\rangle)), \tag{2.26}$$

$$(\tfrac{p}{4}, \tfrac{-i}{\sqrt{2}} \cdot (|00\rangle - |11\rangle)), \tag{2.27}$$

$$(\tfrac{p}{4}, \tfrac{-1}{\sqrt{2}} \cdot (|01\rangle - |10\rangle))\} \tag{2.28}$$

which cannot be represented by a single two-qubit state anymore.

Decoherence errors occur due to the fragile nature of quantum systems (qubits). In practice, this leads to the problem that they can hold information for a limited time only. There are two types of decoherence errors that may appear [104]:

- A qubit in a high-energy state ($|1\rangle$) tends to relax into a low-energy state ($|0\rangle$) or a qubit in a low-energy state ($|0\rangle$) gets energy from its environment and jumps into a high-energy state ($|1\rangle$). This error is called an *amplitude damping error* or *T1 error*.
- In addition, when a qubit interacts with the environment, a phase-flip effect might occur. This leads to a *phase-flip error* or *T2 error*.

Developments in the physical realization of quantum computers show significant improvements in the coherence times of qubits (e.g., in [30, 63]), thereby improving the "lifetime" of qubits. For IBM's ibm_sherbrooke quantum computer the median T1 is at about 294 µs and the median T2 is about 167 µs [63, 102].

Example 2.6 Once again, suppose that the two-qubit register

$$|\varphi'\rangle = \begin{bmatrix} 0 & \tfrac{1}{\sqrt{2}} & \tfrac{1}{\sqrt{2}} & 0 \end{bmatrix}^\top = \frac{1}{\sqrt{2}} \cdot (|01\rangle + |10\rangle) \tag{2.29}$$

from Example 2.2 is affected by an error. More precisely, an amplitude damping error affects qubit q_0 with probability p. This error can be described by applying either

$$\begin{bmatrix} 0 & \sqrt{p} \\ 0 & 0 \end{bmatrix} \text{ or } \begin{bmatrix} 1 & 0 \\ 0 & \sqrt{1-p} \end{bmatrix} \tag{2.30}$$

and then normalizing the state again [82]. Thus, the resulting state mixtures is

$$\{(\tfrac{p}{2}, |00\rangle), (1 - \tfrac{p}{2}, \frac{\sqrt{1-p}}{\sqrt{2-p}} \cdot |01\rangle + \frac{1}{\sqrt{2-p}} \cdot |10\rangle)\}. \tag{2.31}$$

Part II
Foundations

Quantum Circuit Simulation With Decision Diagrams

This book explores the potential of decision diagram-based noise-aware quantum circuit simulation. The starting point for this investigation is the simulation of quantum computers without any consideration of noise. Although this simulation style oversimplifies what is happening in a real quantum computer, it is still an important tool for the analysis of new quantum algorithms and the logical starting point for the realization of noise-aware quantum circuit simulators.

From a mathematical point of view, simulating a quantum circuit on classic hardware is simple and boils down to matrix-vector-multiplication. Accordingly, many state-of-the-art quantum circuit simulators use arrays to represent vectors and matrices, then simulation is performed by matrix-vector-multiplication on these arrays (e.g., [7, 25, 41, 54, 64, 66, 87, 89, 98, 103, 111]). However, these array-based simulators are severely limited by the inherent exponential size of the involved vectors and matrices with respect to the number of simulated qubits.

To overcome the restrictions of array-based quantum circuit simulators, several other simulation approaches have been developed, e.g. based on the *stabilizer formalism* [1, 42], *tensor networks* [14, 73, 74, 86, 108, 109] or *decision diagrams* [2, 44, 79, 80, 85, 94, 105–107, 121]. These simulation styles have their own limitations, e.g., restrictions on the employable quantum operations, exponential growth with respect to the degree of entanglement, or feasibility only for low-depth circuits. This book focuses on the use of decision diagrams for noise-aware quantum circuit simulation.

This chapter (based on [47, 49]) sets the foundation for the upcoming investigation by reviewing and evaluating the current state-of-the-art on decision diagram-based quantum circuit simulation. To this end, the cost of multiplication and addition of decision diagrams—essential operations for quantum circuit simulation—is evaluated first.

T. Grurl et al., *Noise-Aware Quantum Circuit Simulation with Decision Diagrams*,
Synthesis Lectures on Engineering, Science, and Technology,
https://doi.org/10.1007/978-3-031-71036-0_3

Interestingly, while the cost of the two operations strongly correlates with the number of nodes in the involved decision diagrams, the actual cost of multiplication can vary significantly. In contrast, adding decision diagrams depends only on the number of nodes. Next, in a comprehensive case study, a state-of-the-art decision diagram-based quantum circuit simulator is evaluated against an array-based simulator from Atos. The results show that the decision diagram-based solution performs magnitudes faster while using less hardware resources for many quantum algorithms (e.g., for Shor's algorithm to factor integers), thereby confirming the utility of decision diagrams for quantum circuit simulation.

The remainder of this chapter is structured as follows: Sect. 3.1 introduces how quantum circuit simulation and decision diagram-based quantum simulation are conducted from a conceptual point of view. Next, Sect. 3.2 analyzes the cost of multiplying and adding decision diagrams. Then, in Sect. 3.3 the setup of the case study is presented, followed by the presentation and discussion of the results. Finally, the chapter is concluded in Sect. 3.4.

3.1   Quantum Circuit Simulation

This section first reviews how quantum circuit simulation is conducted. Although, this simulation is conceptually simple, it quickly requires huge computational resources due to the exponentially large representation of quantum states and operations. This book considers tackling this complexity using decision diagrams whose principles and utilization are reviewed next.

3.1.1   Array-Based Simulation

The basic concepts reviewed in Sect. 2.1 are sufficient to simulate the execution of *perfect* quantum computers. More precisely, having states and operations represented by one-dimensional and two-dimensional arrays, respectively, the application of operations is simulated by matrix-vector-multiplication, as shown in Example 2.2. This scheme comes with a great potential for parallelization, since every multiplication of a matrix M and a vector V can be split into multiple multiplications and additions, i.e.,

$$\begin{bmatrix} M_{00} & M_{01} \\ M_{10} & M_{11} \end{bmatrix} \cdot \begin{bmatrix} V_0 \\ V_1 \end{bmatrix} = \begin{bmatrix} M_{00} \cdot V_0 + M_{01} \cdot V_1 \\ M_{10} \cdot V_0 + M_{11} \cdot V_1 \end{bmatrix}. \tag{3.1}$$

The outlined decomposition scheme can be repeated recursively, resulting in many intermediate operations that can be performed independently of each other with little synchronization overhead. This conceptually simple approach is the basis for many state-of-the-art quantum circuit simulators such as, e.g., [7, 25, 41, 54, 64, 66, 87, 89, 98, 103, 111]. However, a crucial bottleneck of such a straightforward approach to quantum circuit simulation is the exponential growth of the involved vectors and matrices relative to the number of simulated qubits. This makes the representation of quantum states consisting of more than a few qubits extremely challenging. One might think that considering qubits locally avoids this exponential blow-up, i.e., instead of considering all qubits within a single state at once, each qubit could simply be regarded independently. This is often possible when representing quantum operations (in particular, for elementary operations that use only a few qubits). In these cases, it is not necessary to construct the full operation matrix of size $2^n \times 2^n$ using the Kronecker product. However, this does not extend to the representation of quantum states. Due to *entanglement*—an essential concept of quantum computing—individual qubits may affect each other, making it impossible to represent them individually.

Example 3.1 Consider again the quantum register

$$|\varphi'\rangle = \begin{bmatrix} 0 & \frac{1}{\sqrt{2}} & \frac{1}{\sqrt{2}} & 0 \end{bmatrix}^\top \tag{3.2}$$

from Example 2.2. Measuring the qubit q_0 collapses it to $|0\rangle$ or $|1\rangle$ both with probability $\frac{1}{2}$. However, since q_0 and q_1 are entangled with each other, this measurement also affects q_1. That is, when, e.g., the measurement of q_0 yields $|0\rangle$, then q_1 collapses to the basis state $|1\rangle$ (although not explicitly measured). This shows that, due to entanglement, individual qubits cannot be represented independently of each other.

Overall, although quantum circuit simulation is conceptually simple, this quickly renders the simulation of larger quantum circuits hard, as each additional qubit that is considered effectively doubles the memory required to represent the state vector. Researchers are currently heavily investigating how this exponential complexity can be tackled—leading to different data structures based on the *stabilizer formalism* [1, 42], such as *tensor networks* [14, 73, 74, 86, 108, 109], and *decision diagrams* [2, 44, 79, 80, 85, 94, 105–107, 121]. This book focuses on approaches based on decision diagrams, which are reviewed next.

3.1.2 Decision Diagram-Based Simulation

In order to tackle the exponential complexity of quantum circuit simulation, approaches based on decision diagrams have proven promising. Here, the general idea is to identify data redundancies in the representation of states/operations and to represent them through shared sub-structures. This can result in a very compact representation, which makes it possible to simulate quantum applications that cannot be simulated using a straightforward approach.

More precisely, representing, e.g., a state vector as a decision diagram revolves around recursively splitting the vector into equal-sized subvectors until the subvectors only contain a single element. To illustrate this, consider a quantum register $q_{n-1}, \ldots, q_1, q_0$ composed of n qubits, with q_{n-1} representing the most significant qubit. Then, the first 2^{n-1} entries in the corresponding state vector represent the amplitudes for the basis states where q_{n-1} is $|0\rangle$ and the other entries represent the amplitudes where q_{n-1} is $|1\rangle$. In a decision diagram, this is represented by a node labeled q_{n-1} with two successors, where the left successor points to a node (labeled q_{n-2}) that represents the subvector with amplitudes for basis states with q_{n-1} equal to $|0\rangle$ and the right successor points to a node (also labeled q_{n-2}) that represents the subvector with amplitudes for basis states with q_{n-1} equal to $|1\rangle$.

This scheme is repeated recursively on the subvectors until "subvectors" of size 1 (i.e., complex numbers) remain which are represented as terminal nodes. During this procedure, equivalent subvectors are represented by the same node, which reduces the overall size of the decision diagram. Additionally, instead of having explicit terminal nodes for all amplitudes, edge weights are used to store common factors of the amplitudes, resulting in even more compaction. To reconstruct the amplitude of a specific state, the edge weights along the corresponding path are multiplied. Furthermore, to aid readability of the decision diagram, edge weights of 1 are omitted, and nodes with an incoming edge weight of 0 are represented as 0-stubs, indicating that the amplitudes of all possible states represented by this part of the decision diagram are 0.

Example 3.2 In Fig. 3.1, the quantum state $|\varphi'\rangle$ of Example 2.2 is represented in both the vector and the decision diagram representation. The annotations of the state vector indicate how it is decomposed into the corresponding decision diagram. To reconstruct specific amplitudes from the decision diagram, the edge weights of the corresponding path have to be multiplied. For example, reconstructing the amplitude of the state $|10\rangle$ is done by multiplying the edge weight (see bold lines in Fig. 3.1b) of the root edge ($\frac{1}{\sqrt{2}}$), with the left edge of q_1 (1) and the right edge of q_0 (1), i.e., $\frac{1}{\sqrt{2}} \cdot 1 \cdot 1 = \frac{1}{\sqrt{2}}$.

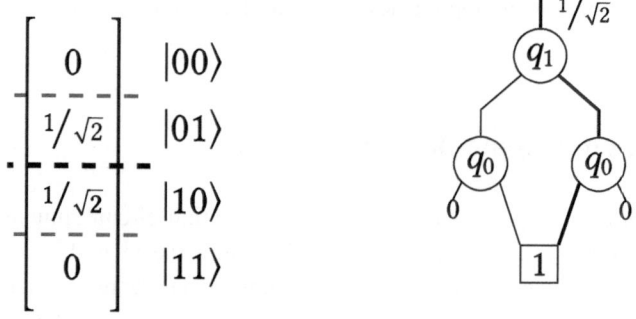

(a) Vector representation of $|\varphi'\rangle$ (b) DD representation of $|\varphi'\rangle$

Fig. 3.1 Decision diagram (DD) representation of a quantum state

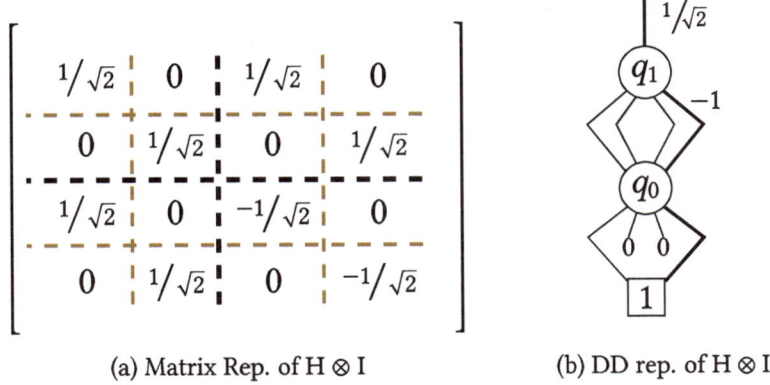

(a) Matrix Rep. of $H \otimes I$ (b) DD rep. of $H \otimes I$

Fig. 3.2 Decision diagram representation of a quantum operation

Matrices representing quantum operations can be decomposed into decision diagrams in a similar fashion to state vectors. However, due to the two-dimensional nature of the matrices, the decomposition process has to be adapted and the matrices are quartered into four equally sized subparts instead of split into two. More precisely, a quantum operation that acts on n qubits, with q_{n-1} representing the most significant qubit, is quartered into four equally sized submatrices. This is represented in a decision diagram by a node labeled q_{n-1} with four successors labeled q_{n-2}: the first represents the submatrix in the upper left corner, the second represents the submatrix in the upper right corner, the third represents the submatrix in the lower left corner, and the fourth represents the submatrix in the lower right corner. The decomposition process is repeated recursively until submatrices of size 1 (i.e., complex numbers) remain. Equivalent submatrices are again represented by the same node, and common factors of encoded (complex) values are extracted and stored as edge weights. To retrieve individual values of the represented matrix, the edge weights along the corresponding path must be multiplied.

Example 3.3 In Fig. 3.2, both the matrix and decision diagram representation of the operation $H \otimes I$ from Example 2.2 are provided. Once again, the annotations of the matrix representation indicate how the matrix is decomposed into the resulting decision diagram. Analogously to the vector decomposition, the reconstruction of specific matrix elements from the decision diagram representation is done by multiplying the edge weights along the corresponding path. For example, the lower right element can be reconstructed by multiplying (see bold lines in figure) the root edge ($\frac{1}{\sqrt{2}}$), with the fourth edge of q_1 (-1) and the first edge of q_1 (1), i.e., $\frac{1}{\sqrt{2}} \cdot -1 \cdot 1 = \frac{-1}{\sqrt{2}}$.

Having a representation of quantum states and operations through decision diagrams, simulation is conducted by multiplying operations onto states. Yet, due to the different representation, multiplications must be decomposed with respect to the most significant qubit, which can be illustrated using matrix-vector-multiplication: Consider a quantum register $|\psi\rangle = |q_{n-1}, \ldots, q_1, q_0\rangle$ of n qubits, where q_0 represents the most significant qubit and a unitary quantum operation U of size $2^n \times 2^n$. Multiplying the operation U onto the state $|\psi\rangle$, is done by splitting $|\psi\rangle$ into two and U into four equally sized parts, leading to two subvectors of size 2^{n-1} and four submatrices of size $2^{n-1} \times 2^{n-1}$. The subvectors and submatrices represent the outgoing edges of the root nodes of the decision diagrams representing $|\psi\rangle$ and U. The process is repeated recursively until matrices of size 2×2 and vectors of size 2 remain, which are multiplied. From the resulting values (i.e., amplitudes), the edge weights are calculated by extracting common factors, and the new decision diagram is constructed.

This process can be illustrated using matrix-vector notation. The multiplication of the decision diagrams U and $|\psi\rangle$ then becomes

$$U \cdot |\psi\rangle = \begin{bmatrix} U_0 & U_1 \\ U_2 & U_3 \end{bmatrix} \cdot \begin{bmatrix} \psi_0 \\ \psi_1 \end{bmatrix} = \begin{bmatrix} U_0 \cdot \psi_0 + U_1 \cdot \psi_1 \\ U_2 \cdot \psi_0 + U_3 \cdot \psi_1 \end{bmatrix}, \tag{3.3}$$

With U_0 representing the first outgoing edge of U, U_1 representing the second outgoing edge of U, U_2 representing the third outgoing edge of U, U_3 representing the fourth outgoing edge of U, ψ_0 represents the first outgoing edge of $|\psi\rangle$ and ψ_1 represents the second outgoing edge of $|\psi\rangle$. The resulting decision diagram consists of a top node with the left successor representing $U_0 \cdot \psi_0 + U_1 \cdot \psi_1$ and the right successor representing $U_2 \cdot \psi_0 + U_3 \cdot \psi_1$. These successors are decomposed recursively in a similar fashion until only operations on complex numbers remain, which can be directly multiplied. Afterwards, the obtained subresults are added together accordingly in order to obtain the resulting state vector.

To add decision diagrams, the operation has to be decomposed again with respect to the most significant qubit. Here, again, the vector notation is helpful for illustration. Calculating the sum of the two quantum states $|\psi\rangle$ and $|\phi\rangle$ (which are both represented as decision diagrams) can be written as

$$|\psi\rangle + |\phi\rangle = \begin{bmatrix} \psi_0 \\ \psi_1 \end{bmatrix} + \begin{bmatrix} \phi_0 \\ \phi_1 \end{bmatrix} = \begin{bmatrix} \psi_0 + \phi_0 \\ \psi_1 + \phi_1 \end{bmatrix}, \tag{3.4}$$

with ψ_0 representing the first outgoing edge of $|\psi\rangle$, ψ_1 representing the second outgoing edge of $|\psi\rangle$, ϕ_0 representing the first outgoing edge of $|\phi\rangle$, and ϕ_1 representing the second outgoing edge of $|\phi\rangle$. As with multiplication, the left successor is represented by the upper part $\psi_0 + \phi_0$ and the right successor is represented by the lower part $\psi_1 + \phi_1$. The successors are recursively decomposed until only operations on complex numbers remain, that are added.

Hence, decision diagram-based simulation mainly involves recursive traversals of the involved decision diagrams. On top of that, further optimizations are possible with respect to the precision of the simulation [120], approximation [56, 118] or qubit order [35, 55, 76–78].

Overall, using those structures often allows one to gain representations of states and operations which are much more compact than the exponentially large vectors and matrices. Since matrix-vector-multiplication (and, hence, quantum circuit simulation) only has a complexity that is linear with respect to the number of nodes in the corresponding decision diagrams, this allows for a substantially more efficient simulation in these cases.

3.2 Cost of Operations on Decision Diagrams

As discussed in Sect. 3.1.1, the simulation of quantum circuits on classical hardware boils down to a series of matrix-vector-multiplications. Thus, the cost of performing multiplication and addition (the latter being a suboperation of multiplying decision diagrams) is essential for fast quantum circuit simulation. Therefore, this section sheds some light on this matter by evaluating the cost of multiplication and addition of decision diagrams.

3.2.1 Multiplication on Decision Diagrams

Multiplication is arguably the most important operation for quantum circuit simulation. Hence, it has already been discussed in detail in related work, such as [121]. However, two aspects have not yet been taken into account so far: (1) the position of the modified qubit within the decision diagram representation and (2) the type of the applied quantum operation. These aspects can have a significant impact on the cost of applying operations and thus conducting quantum circuit simulation.

To illustrate how the position of the modified qubit affects the cost of applying operations, consider the following example:

Example 3.4 Consider the three-qubit state

$$\phi = \frac{1}{\sqrt{2}} \begin{bmatrix} 1\,0\,0\,0\,0\,0\,0\,1 \end{bmatrix}^{\top} \tag{3.5}$$

whose decision diagram representation is provided in Fig. 3.3a. An X operation is applied to the qubit represented by the root node (q_2) and the qubit farthest away from it (q_0), respectively:

- Applying an X operation to q_2 results in swapping the two outgoing edges of the node q_2, as shown in Fig. 3.3b. Since q_2 is located at the root of the decision diagram, only *one* node has to be visited to apply the operation.
- Analogously, applying the X operation to q_0, requires swapping the outgoing edges of all nodes labeled q_0, resulting in the decision diagram shown in Fig. 3.3c. However, due to the tree-like structure, accessing the nodes q_0 requires a full traversal of the decision diagram.

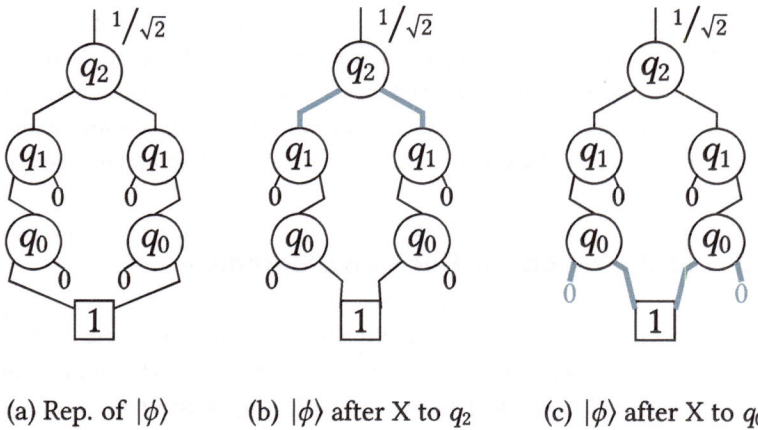

(a) Rep. of $|\phi\rangle$ (b) $|\phi\rangle$ after X to q_2 (c) $|\phi\rangle$ after X to q_0

Fig. 3.3 Modifications of state $|\phi\rangle$, when applying an X operation

Moreover, q_0 is represented by two nodes. Thus, applying the operations requires visiting *five* nodes.

So, although in both cases the same operation is applied, the cost of doing so can vary drastically depending on the target qubit. While only one node has to be visited in order to apply the operation to q_2, modifying q_0 requires to traverse the *entire* decision diagram.

In addition to the position of the modified qubit, the type of operation also affects the cost of applying it. This stems from the fact that many quantum operations, such as X, Z, Y, or T only modify the target qubit by some factor and/or by swapping of sub-trees. In these cases, the compressed data structure of decision diagrams can be explicitly exploited. More precisely, recall that the amplitudes of the quantum state are encoded within the decision diagram via multiplication. Consequently, modifying a qubit by some factor does not require decompressing any amplitudes; instead, it can be done by directly modifying the edge weights of the nodes representing the specific qubit. Therefore, the cost of applying such operations becomes the cost of accessing all nodes representing the qubit the operation is applied to. Given that decision diagrams encode the quantum state in a tree-like structure, qubits closer to the root can be accessed faster—making applying such operations potentially extremely fast.

Unfortunately, this effect cannot be exploited for all quantum operations. Examples of operations where this effect cannot be exploited are H, RX(θ) or RY(θ). Applying those operations not only requires accessing the nodes representing the target qubit but also access to all subnodes. Thus, it is not relevant how deep the target qubit is within the decision diagram, since it has to be traversed anyway.

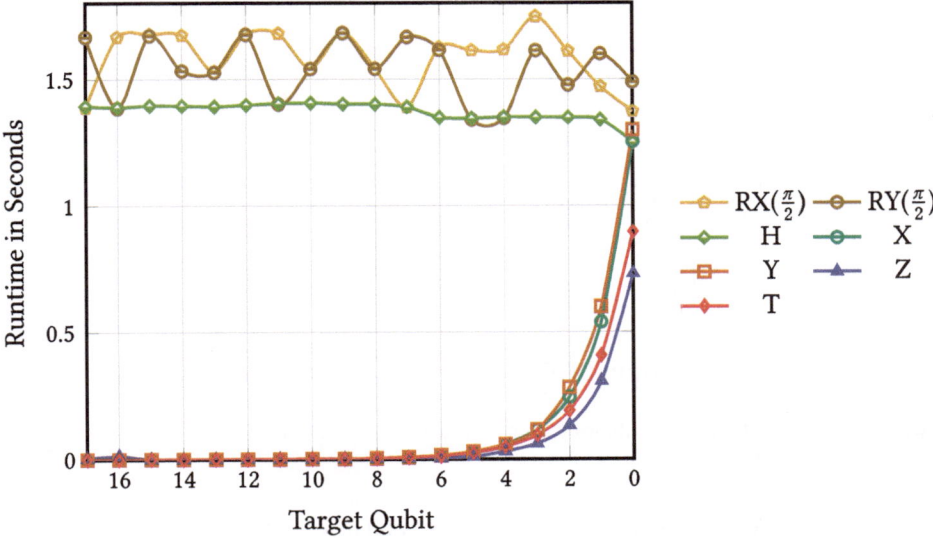

Fig. 3.4 Avg. executing time for applying operation depending on the target qubit

Example 3.5 To illustrate this effect, the average time it took to apply specific quantum operations to an 18-qubit quantum system, depending on the target qubit, is tracked. The operation is always applied to the same quantum state, which is prepared in such a way as to contain few redundancies, using Google's supremacy circuit [15]. The results are shown in Fig. 3.4. The Y-axis shows the average (avg.) execution time of the operation, while the X-axis indicates how deep the target qubit is within the decision diagram, with 17 denoting that the target qubit is directly represented by the root node and 0 denoting that the target qubit is represented by nodes that are directly above the terminal node. The figure clearly shows that, when an H, $RX(\frac{\pi}{2})$, or $RY(\frac{\pi}{2})$ operation is applied, the cost is not affected by the position of the target qubit. In contrast, when an X, Z, Y, or T operation is applied, the cost depends on how deep the target qubit is within the decision diagram.

3.2.2   Addition on Decision Diagrams

Multiplication is arguably the most important operation during quantum circuit simulation. Addition, however, is also important, as it is a necessary subroutine when multiplying decision diagrams. This makes adding decision diagrams an operation worth considering. In fact, adding decision diagrams is not straightforward, as the following example shows:

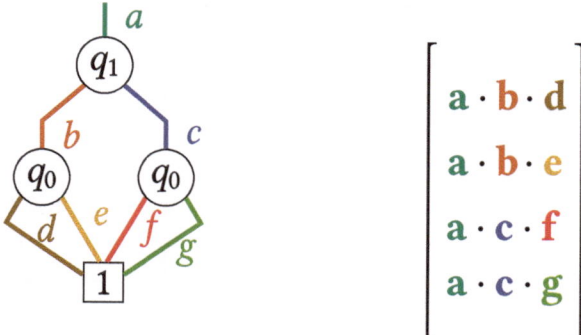

(a) Generic two-qubit DD   (b) Generic two-qubit state vector

Fig. 3.5 Decomposition of amplitudes within a decision diagram (DD)

Example 3.6 Adding two decision diagrams is similar to adding vectors (or matrices), i.e., all values sharing the same index are accessed and added together. However, this process poses a challenge for decision diagrams, since the values are encoded within the tree structure. Figure 3.5 illustrates how a generic two-qubit state vector is decomposed into a decision diagram (see Fig. 3.5a) and how the individual edge weights relate to the amplitudes of the corresponding state vector (see Fig. 3.5b). Adding two decision diagrams therefore becomes a full recursive traversal of them both, so that all amplitudes can be restored and added together. Additionally, after the amplitudes have been added, they have to be again decomposed into the decision diagram representation.

Restoring all amplitudes encoded into a decision diagram is inherently slow. Other decision diagram decomposition schemes are possible, which allow adding decision diagrams without having to decompress the stored values. However, doing so without making the multiplication of edges slower is problematic. Current decomposition schemes for quantum decision diagrams are optimized for multiplication of edges—since this is the more important operation when decision diagrams are multiplied.

Nevertheless, two shortcuts are possible compared to the naive approach described in Example 3.6: Firstly, when a zero edge is encountered during the recursive traversal in one of the decision diagrams, all amplitudes along this path are zero. Therefore, the new amplitudes are simply those stored in the remaining decision diagram. Secondly, when two identical (sub-)edges are added, one edge can instead be multiplied by a factor of two. Finding such duplicate paths is straightforward, since, due to the compressed data structure, identical values are represented by the same node within decision diagrams.

3.3   Case Study

To shed light on the strengths and weaknesses of quantum decision diagrams. The runtime and memory requirements of quantum decision diagrams are evaluated for different quantum applications. Additionally, to establish a baseline, all quantum applications are also simulated with a state-of-the-art array-based quantum circuit simulator. The setup of the case study is described next, followed by the presentation and discussion of the results.

3.3.1   Setup

As a state-of-the-art representative of an array-based simulator, the *Linalg* simulator from the commercial *Atos Quantum Learning Machine* (QLM) [7] is used. The QLM is an environment for quantum computing, which has been developed since 2016. Among other things, it comes with its own quantum simulators, circuit optimizers, dedicated hardware, and a customized Linux kernel. In 2018, the QLM was the first simulator that allowed quantum circuits to be simulated under consideration of noise effects [6]. The particular QLM instance used utilizes 96 cores running at a clock frequency of 2.2 GHz and 1.5 TB of RAM.

As a state-of-the-art representative of a decision diagram-based simulator, the implementation taken from [119] (based on the principles of [121]) is used. To allow a fair comparison, the decision diagram-based simulator is ported to the QLM using Docker [75]. Docker is used since its virtualization overhead is negligible [36]. Therefore, both simulators are running on the same hardware and have the same resources available. In practice, however, the decision diagram-based simulator currently does not make use of parallel executions (using concurrency for decision diagram-based simulation is still an active field of research [57]), while the array-based simulator uses the full potential of the hardware resources (this way it has 96 times more processor power at its disposal).

Finally, in order to properly evaluate the performance of both simulators, different types of benchmark circuits have been considered. This includes benchmarks representing certain quantum characteristics (namely the *Entanglement*-benchmarks, which construct the GHZ state [82]), benchmarks realizing certain quantum algorithms (namely the quantum Fourier transform (QFT) [82], Grover's algorithm for database search [46] and Shor's factorization algorithm [96]), as well as certain random circuits designed by Google in their development toward *quantum advantage* [15], which are engineered to be exceptionally hard to simulate classically. All simulations were conducted using a variable number of qubits, allowing for a detailed evaluation of the scalability of both simulation approaches.

3.3.2 Obtained Results

All obtained results are summarized in Figs. 3.6, 3.7, 3.8, 3.9 and 3.10. The respective
plots provide the runtime (left side) and the memory consumption (right side) for both
the array-based and decision diagram-based (DD-based) approach, and for each benchmark
(scaling with respect to the number of qubits). Simulations that could not be completed within
one hour are omitted (i.e., the respective curve stops at the largest number of qubits that
could be simulated within this time). Subsequently, the obtained results for each benchmark
are discussed separately. This provides the basis of the conclusions which are summarized
afterwards.

The data is presented in the form of logarithmic plots, since they give a more immediate
impression of how the simulators behave for specific benchmarks. In the remainder of this
section, the behavior of both simulators is described for each benchmark.

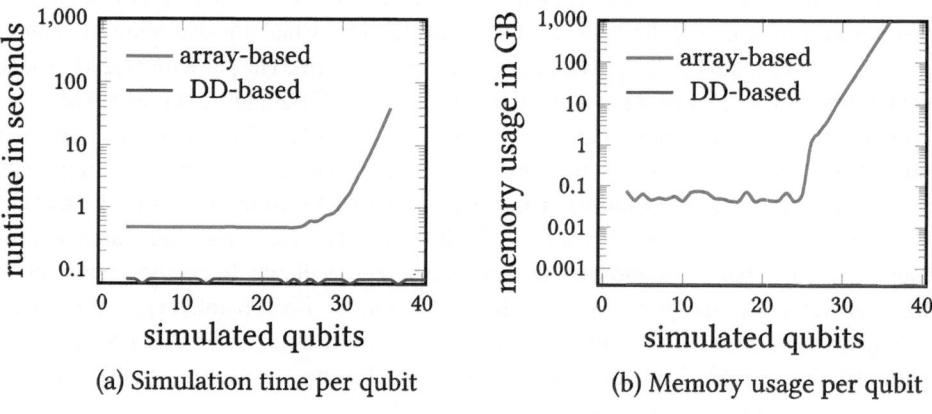

Fig. 3.6 Entanglement circuit simulation

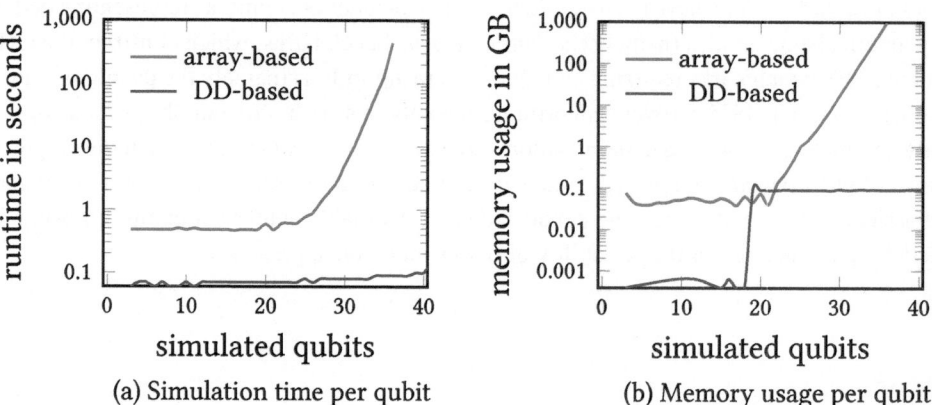

Fig. 3.7 Quantum Fourier transform circuit simulation

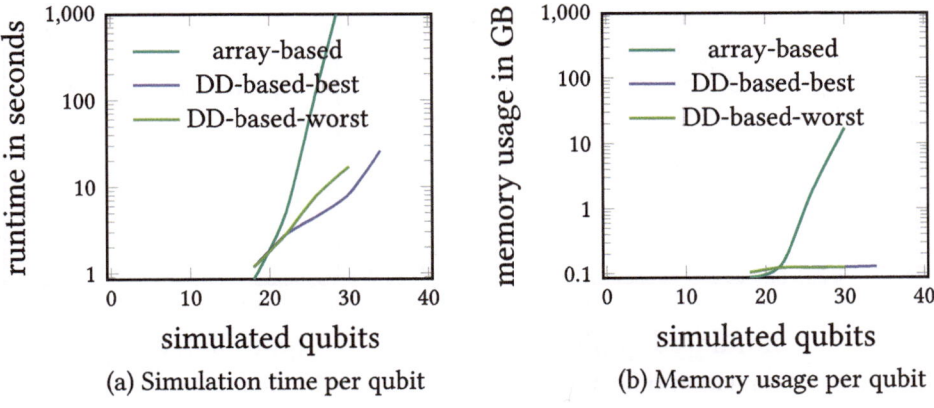

(a) Simulation time per qubit

(b) Memory usage per qubit

Fig. 3.8 Shor's algorithm circuit simulation

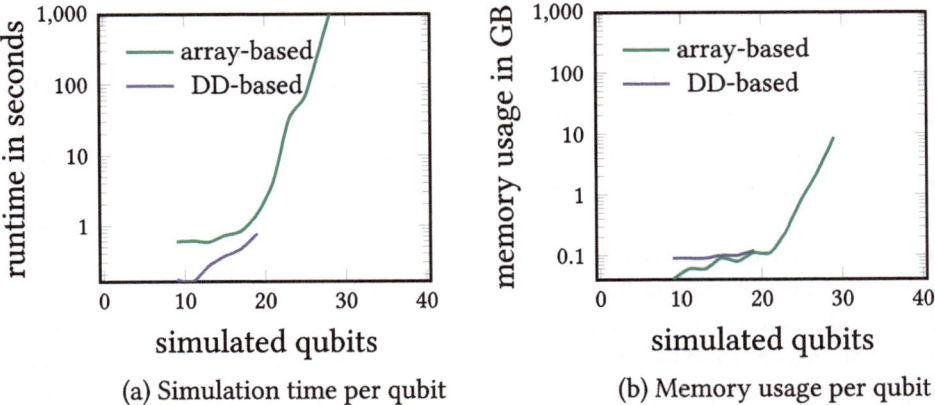

(a) Simulation time per qubit

(b) Memory usage per qubit

Fig. 3.9 Grover's algorithm circuit simulation

Considering the Entanglement-benchmarks (see Fig. 3.6) and the QFT-benchmarks (see Fig. 3.7), almost identical performance can be observed: The runtime and memory consumption of the array-based simulator grows exponentially with respect to the number of simulated qubits. In contrast, the decision diagram-based simulator shows linear growth in runtime and memory consumption. This limits the array-based approach to a maximum of 36 qubits, while the decision diagram-based approach can be scaled to hundreds of qubits in these cases.

Considering Shor's algorithm (see Fig. 3.8), fluctuations in the runtime of the decision diagram-based simulator can be observed. On the one hand, this is due to the fact that multiple instances of Shor's algorithm with the same number of qubits but different integers to be factored were used during the evaluations. Naturally, the factorization of different integers leads to different computations during the simulation, which in turn result in decision

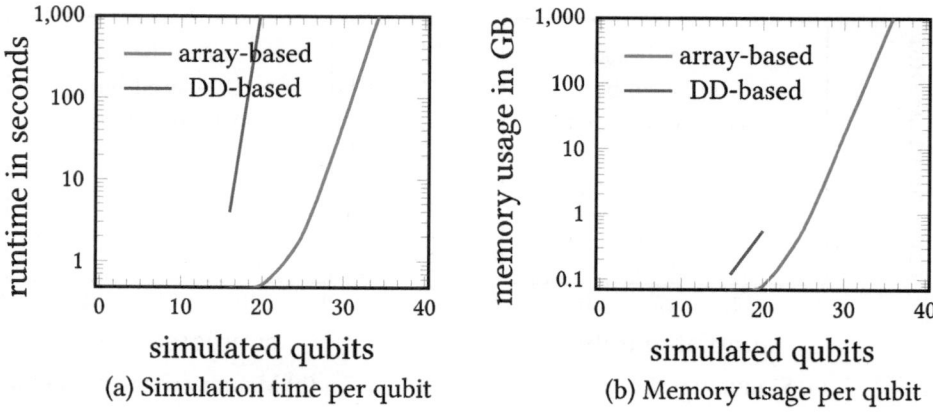

(a) Simulation time per qubit   (b) Memory usage per qubit

Fig. 3.10 Quantum supremacy circuit simulation

diagrams of varying sizes. On the other hand, the implementation used in the evaluations contains intermediate measurements. While the array-based simulator is not affected by these measurements (it still stores the entire state vector), the decision diagram-based simulator exploits the measurement collapse of the quantum state to achieve compaction. Since the measurement process is inherently probabilistic, its execution during the simulation depends on a given random seed—further explaining the varying runtimes. To properly reflect that in the results, both, the best and worst, runs for the decision diagram-based simulator are plotted in Fig. 3.8. Independently of this, an exponential growth can be observed in all cases. The decision diagram-based simulator manages to simulate circuits composed of 30 qubits in its worst runs, which is also the limit of the array-based simulator. In the best case, the decision diagram-based simulator can simulate Shor's algorithm with up to 34 qubits, making the decision diagram-based approach the better choice to simulate this important algorithm.

For Grover's algorithm (see Fig. 3.9), both simulators exhibit exponential growth as well. But here, the array-based simulator manages to simulate more qubits than the decision diagram-based simulator (namely, up to 29 qubits in the case of array-based simulation and up to 19 qubits in the case of decision diagram-based simulation).

Finally, considering Google's quantum supremacy benchmarks (see Fig. 3.10), the array-based simulator performs significantly better than the decision diagram-based simulator[1]. This difference arises due to the specific structure of these benchmarks. They are specifically designed in a way that almost no redundancies can be exploited. Thus, the overhead in decision diagram-based simulation, incurred by deriving more compact representations by extracting common factors and sharing nodes, does not pay off in these cases. Instead, the array-based approach keeps following its straightforward scheme, which allows for simulation up to a limit similar to the other benchmarks.

---

[1] In the plots, results are only provided for 16, 20, 25, 30 and 36 qubits, where the decision diagram-based simulator was only able to simulate the first two instances.

3.4 Conclusion

This book explores the potential of decision diagrams for noise-aware quantum circuit simulation. The logical starting point of this investigation is the decision diagram-based simulation of *perfect* quantum circuits, for which quantum decision diagrams are already used successfully. Thus, this chapter reviewed and evaluated the state-of-the-art of decision diagram-based quantum circuit simulation and sheds light on their strengths and weaknesses.

To this end, the cost of multiplication and addition of decision diagrams—essential operations for quantum circuit simulation—was evaluated. While the cost of both operations strongly correlates with the number of nodes in the decision diagrams involved, the actual cost of multiplication can still vary considerably depending on the structure of the involved decision diagrams. This is in stark contrast to adding decision diagrams, which depends only on the number of nodes.

Afterwards, the performance of decision diagram-based quantum circuit simulation was compared against an array-based quantum circuit simulator from Atos, in the course of an extensive case study. The results show that the runtime and memory footprint of the array-based simulator always grows exponentially with the number of simulated qubits and can be reliably predicted before execution. In contrast, by using a potentially more compact representation, the decision diagram-based solution was often able to simulate more qubits, e.g., for Shor's algorithm to factor integers.

Overall, the utility of decision diagrams for quantum circuit simulation can be confirmed and the evaluation of decision diagram-based quantum circuit simulation serves as a starting point for the remainder of this book.

Decision Diagrams for Density Matrices

4

The previous chapter showed how decision diagrams can often keep the runtime and memory requirements of quantum circuit simulation in check by exploiting redundancies within their classical description. However, all developments on quantum decision diagrams so far have mainly been focused on simulating *perfect* quantum computers only. This is in contrast to how real quantum computers behave, as they are heavily affected by frequent noise effects caused by the fragile nature of quantum mechanical effects [88]. Considering these errors during quantum circuit simulation is necessary in order to gain an accurate picture of what would happen when a circuit is executed on a real quantum computer.

Fortunately, those noise effects are well understood and hence, according descriptions in terms of the so-called *mixed quantum states* described by means of *density matrices* exist [82]. But, although they allow one to properly represent a quantum state including possible noise effects, they are substantially larger than the (already exponential) pure quantum state representations. As a result, most existing quantum circuit simulators either do not support a noise-aware consideration of quantum states yet (e.g., [79, 85, 87, 103, 108, 121]) or are heavily limited in their efficiency and scalability (e.g., [7, 24, 25, 41, 54, 64, 66, 73, 89, 98, 109, 111]). This urgently motivates further research towards more optimized density matrix representations.

In this chapter (based on [51]), an approach towards optimized density matrix representations based on decision diagrams is proposed. Although decision diagrams can directly represent density matrices, the Hermitian property of density matrices has not been exploited yet. However, doing this allows for the utilization of redundancies that have not been caught before (leading to a more compact representation in many cases). This chapter illustrates and uses this untapped potential in an optimized decision diagram representation for density matrices. Afterwards, the magnitude of the resulting potential is discussed and empirically

© The Author(s), under exclusive license to Springer Nature Switzerland AG 2025
T. Grurl et al., *Noise-Aware Quantum Circuit Simulation with Decision Diagrams*,
Synthesis Lectures on Engineering, Science, and Technology,
https://doi.org/10.1007/978-3-031-71036-0_4

confirmed using quantum circuit simulation. The evaluation shows that the improved density matrix representation is never larger than the established decision diagram matrix representation and up to 50% more compact. Furthermore, the smaller size of the proposed decision diagrams results in runtime improvements for quantum circuit simulation of up to 53%.

The remainder of this chapter is structured as follows: Sect. 4.1 discusses the shortcomings of current representations of density matrices in terms of decision diagrams and illustrates the potential that is yet untapped. Section 4.2 then presents how this leads to an optimized density matrix representation. Afterwards, the magnitude of the unveiled potential is discussed in Sect. 4.3 and empirical results confirming this are presented in Sect. 4.4. Finally, the chapter is concluded in Sect. 4.5.

4.1   Motivation and General Idea

In this section, the general idea behind an improved representation of density matrices is proposed. To this end, density matrices are introduced, and it is shown how these structures can be used to represent state mixtures. Next, the concept of recursively decomposing matrices into smaller structures is presented, the basis for current matrix representation with quantum decision diagrams. Afterwards, the shortcomings when working with density matrices are illustrated, and the proposed decomposition scheme is presented, which additionally takes characteristics of them into account.

4.1.1   Density Matrix Representation

As reviewed in Sect. 2.2, errors probabilistically affect qubits and leave a pure state in a state mixture. Since each possible state in the mixture can once again be affected by errors, the number of possible states increases exponentially with each error, and thus it quickly becomes infeasible to track all possible quantum states separately. Fortunately, quantum mechanics can describe such state mixtures with a single structure, using so-called *density matrices* [82].

Definition 4.1 Let $\{(p_i, |\psi_i\rangle)|1 \leq i \leq k\}$ be a state ensemble of k pure states as defined in Definition 2.5. The corresponding *density matrix* is defined as

$$\rho = \sum_{i=1}^{k} p_i |\psi_i\rangle\langle\psi_i| \tag{4.1}$$

Example 4.1 Suppose the quantum register

$$|\varphi'\rangle = \begin{bmatrix} 0 & 1/\sqrt{2} & 1/\sqrt{2} & 0 \end{bmatrix}^{\top} \tag{4.2}$$

from Example 2.2 is affected by an amplitude damping error in q_0 with 2% probability ($p = 0.02$) (analogous to Example 2.6), resulting in the state mixture

$$\{(0.01, |00\rangle), (0.99, 0.703526 \cdot |01\rangle + 0.710669 \cdot |10\rangle)\}. \quad (4.3)$$

The density matrix ρ' representing this state mixture is given by

$$0.01 \cdot \begin{bmatrix} 1 \\ 0 \\ 0 \\ 0 \end{bmatrix} \cdot \begin{bmatrix} 1 & 0 & 0 & 0 \end{bmatrix} + 0.99 \cdot \begin{bmatrix} 0 \\ 0.703526 \\ 0.710669 \\ 0 \end{bmatrix} \cdot \begin{bmatrix} 0 & 0.703526 & 0.710669 & 0 \end{bmatrix} = \quad (4.4)$$

$$\underbrace{\begin{bmatrix} \mathbf{0.01} & 0 & 0 & 0 \\ 0 & \mathbf{0.49} & 0.494975 & 0 \\ 0 & 0.494975 & \mathbf{0.5} & 0 \\ 0 & 0 & 0 & \mathbf{0} \end{bmatrix}}_{\rho'}. \quad (4.5)$$

Analogous to the vector representation, the density matrix contains the probabilities of measuring specific basis states. Now, however, the probabilities are reflected in the diagonal elements (written in bold) of the matrix. More precisely, the diagonal entries from the first element on the upper left to the last element on the lower right represent the probabilities to measure $|00\rangle$, $|01\rangle$, $|10\rangle$, and $|11\rangle$, respectively. Hence, measuring this state would yield $|10\rangle$ with probability 0.5. Due to the amplitude damping error the probability for measuring $|01\rangle$ dropped by 2% to 0.49 and now there is a probability of 0.01 for measuring $|00\rangle$.

Since quantum states are now represented by density matrices (rather than vectors), obviously also the application of quantum operations (represented by matrices thus far) needs to be adjusted. Instead of a matrix-vector-multiplication, now two matrix-matrix-multiplications are required as shown in Definition 4.2.

Definition 4.2 Applying an operation specified by the unitary matrix U to a quantum system given by the density matrix ρ yields the density matrix

$$\rho' = U \cdot \rho \cdot U^\dagger. \quad (4.6)$$

Example 4.2 Consider again the state ρ' from Example 4.1. A CNOT operation with q_0 as the target and q_1 as the control qubit, would be applied to ρ' like so,

$$
\underbrace{\begin{bmatrix} 1 & 0 & 0 & 0 \\ 0 & 1 & 0 & 0 \\ 0 & 0 & 0 & 1 \\ 0 & 0 & 1 & 0 \end{bmatrix}}_{\text{CNOT}} \cdot \underbrace{\begin{bmatrix} 0.01 & 0 & 0 & 0 \\ 0 & 0.49 & 0.494975 & 0 \\ 0 & 0.494975 & 0.5 & 0 \\ 0 & 0 & 0 & 0 \end{bmatrix}}_{\rho'} \cdot \underbrace{\begin{bmatrix} 1 & 0 & 0 & 0 \\ 0 & 1 & 0 & 0 \\ 0 & 0 & 0 & 1 \\ 0 & 0 & 1 & 0 \end{bmatrix}}_{\text{CNOT}^\dagger} = \tag{4.7}
$$

$$
\underbrace{\begin{bmatrix} 0.01 & 0 & 0 & 0 \\ 0 & 0.49 & 0 & 0.494975 \\ 0 & 0 & 0 & 0 \\ 0 & 0.494975 & 0 & 0.5 \end{bmatrix}}_{\rho''}. \tag{4.8}
$$

Measuring ρ'' would yield $|00\rangle$ with probability 0.01, $|01\rangle$ with probability 0.49, and $|11\rangle$ with probability 0.5.

Using the formalism introduced in this section, mixed states can be fully considered in quantum circuit simulators. However, the basic problem of exponentially sized states description still remains and, even, gets worse with density matrices, as they are substantially larger than pure states. Thus, handling those structures becomes an important problem.

4.1.2   Decomposition of Density Matrices

Current Decomposition of Density Matrices

A major challenge when working with quantum states is that they grow exponentially in size with each tracked qubit. As shown in Chap. 3, decision diagrams can often keep resource requirements in check by recursively decomposing the state description and then exploiting redundancies within the description for a more compact representation. Although, thus far, most work on decision diagrams has been focused on working with pure quantum states, current schemes can be readily employed for decomposing density matrices by reusing the decomposition process for matrices.

Example 4.3 To illustrate the idea, consider the density matrix $\rho^\star$ in Fig. 4.1. The decomposition process revolves around recursively quartering the matrix and checking for redundancies within the submatrices. That is, $\rho^\star$ is quartered (as indicated by the dashed black lines), and the submatrices are checked for redundancies. Since all submatrices are different, no redundancies are determined and the decomposition process is repeated on all submatrices (as indicated by the dashed blue lines). In this step redundant submatrices are identified, which are framed in red and orange. Note that although the submatrices

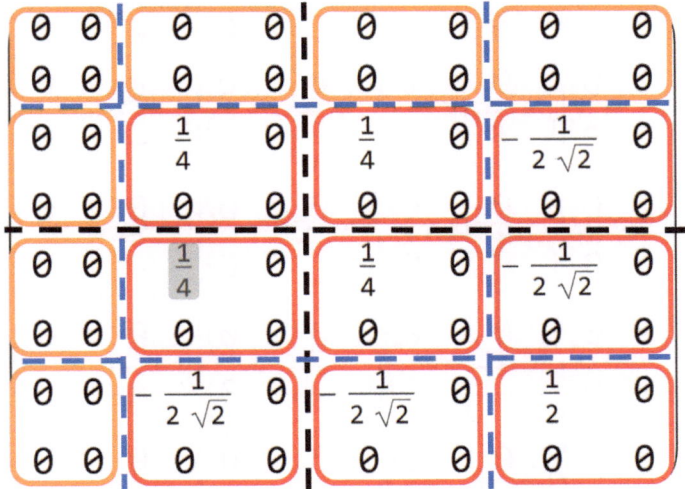

Fig. 4.1 Established matrix decomposition applied to density matrices

$$\begin{bmatrix} \frac{1}{4} & 0 \\ 0 & 0 \end{bmatrix} \text{ and } \begin{bmatrix} -\frac{1}{2\sqrt{2}} & 0 \\ 0 & 0 \end{bmatrix} \tag{4.9}$$

are different, they are multiples of each other and can therefore be represented by the same node using edge weights, as illustrated later in this chapter. The submatrix containing only zeros (framed in orange) is discarded, and the decomposition process is only repeated for the submatrix framed in red. Quartering this submatrix results in complex numbers only—terminating the decomposition process.

After determining those redundancies, the matrix is represented using a decision diagram (as introduced in Sect. 3.1.2), where the redundant structures are only stored once. As illustrated in the previous chapter, this yields descriptions that are often much more compact than straightforward representations.

However, this decomposition scheme was originally developed for representing *pure* quantum state vectors and quantum operations, i.e., representations that do not consider the noise effects as reviewed above in Sect. 2.2. As shown next, this leaves potential which would allow for a more optimized representation of density matrices and hence, quantum state mixtures.

Proposed Decomposition of Density Matrices

While density matrices are substantially larger than state vectors, they also have the nice property that they are Hermitian, i.e., each density matrix M satisfies $M = M^\dagger$. Therefore, each off-diagonal element of a density matrix is mirrored (in a transposed and complex conjugated fashion) to the element on the other side of the diagonal. Thus, up to half of

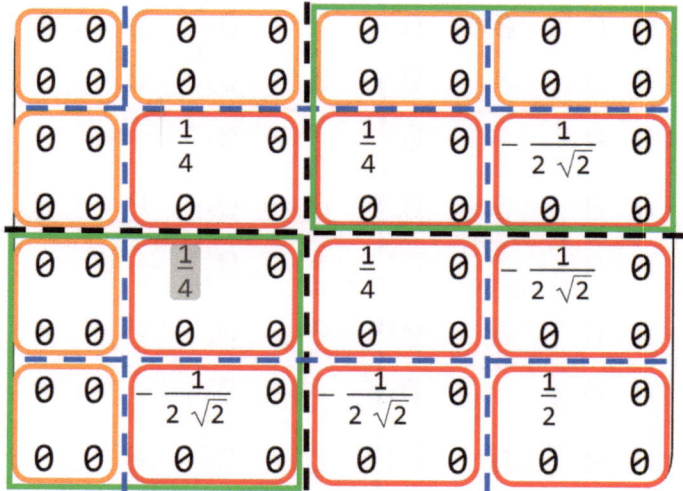

Fig. 4.2 Extended decomposition scheme optimized for density matrices

the elements within the density matrix are redundant. Current (established) decomposition schemes for quantum objects (such as reviewed above and illustrated in Fig. 4.1) do not exploit this potential.

In order to exploit this characteristic, the "mirrored" parts of the matrix are set equal. More precisely, the decomposition scheme described above can be extended in such a way that the matrix elements below the diagonal are discarded and only elements on and above the matrix diagonal are stored. To illustrate the idea, consider the following example:

Example 4.4 Consider again $\rho^\star$ from Example 4.3 and the current (established) decomposition scheme sketched in Fig. 4.1. The newly proposed scheme is illustrated in Fig. 4.2. First, the matrix is again quartered into four submatrices (indicated by the black dashed lines). Although all submatrices appear different, the Hermitian property of density matrices ensures that the upper right and lower left submatrices (framed in green) are just the complex conjugate transpose of each other. Therefore, one of the two submatrices can be discarded without losing any information (without loss of generality, always the lower left submatrix is discarded). Then, the decomposition step is repeated on the remaining three submatrices (as indicated by the blue dashed lines). Here, two identical submatrices can be identified, which are framed in orange and red (the same that could be identified redundantly with the established decomposition scheme reviewed in the previous section). The decomposition process is only continued for the submatrix framed in red, and the process is terminated after complex numbers are reached.

Obviously, exploiting the Hermitian characteristic allows for much more potential in exploiting redundancies and, by this, determining much more compact representations. However, to properly use such an optimized representation and to apply quantum circuit operations on corresponding state representations, some care has to be taken when reconstructing the original matrix elements. This leads to an extended type of decision diagram which is introduced next.

4.2   Optimized Representation of Density Matrices

This section describes how, using the ideas sketched above, optimized decision diagrams representing density matrices can be constructed. Afterwards, it is also presented how quantum operations are applied to the resulting decision diagrams and how this affects their structure.

4.2.1   Representation

As described above, the key strength of decision diagrams is in determining and exploiting redundancies within structures. They are constructed from matrices by following the decomposition process presented above and reflecting this process in a graph-based representation, as presented in Sect. 3.1.2. To recap, consider a density matrix representing a quantum register composed of n qubits with q_{n-1} representing the most significant qubit. During the decomposition process, this density matrix is recursively quartered until submatrices of size 1 (i.e., complex numbers) remain. The first time the matrix is quartered, this is represented in a decision diagram with a node labeled q_{n-1} with four successor nodes that represent the individual submatrices. This process is recursively repeated for $q_{n-2}, q_{n-3}, \ldots, q_0$ until complex numbers remain, which are then connected to a so-called terminal node. Redundant submatrices are represented by the same node (called *shared node*) and common factors within the amplitudes are stored as edge weights, for a more compact representation. Individual elements of the density matrix can be reconstructed from the decision diagram by multiplying the edge weights along the corresponding path.

Example 4.5 Figure 4.3a depicts the decision diagram representation of $\rho^\star$ resulting from the established decomposition scheme. Following established conventions, edge weights of 1 are omitted and nodes with an incoming edge weight of 0 are represented as 0-stubs— indicating that matrix values of all possible states represented by this part of the decision diagram are 0. As can be seen, the redundancies discussed above in Example 4.3 and framed in orange and red in Fig. 4.1 are accordingly reflected in the decision diagram. To obtain the value of a matrix entry from this decision diagram, the edge weights of the corresponding path must be multiplied. For example, to reconstruct the value $\frac{1}{4}$ from the lower left part of

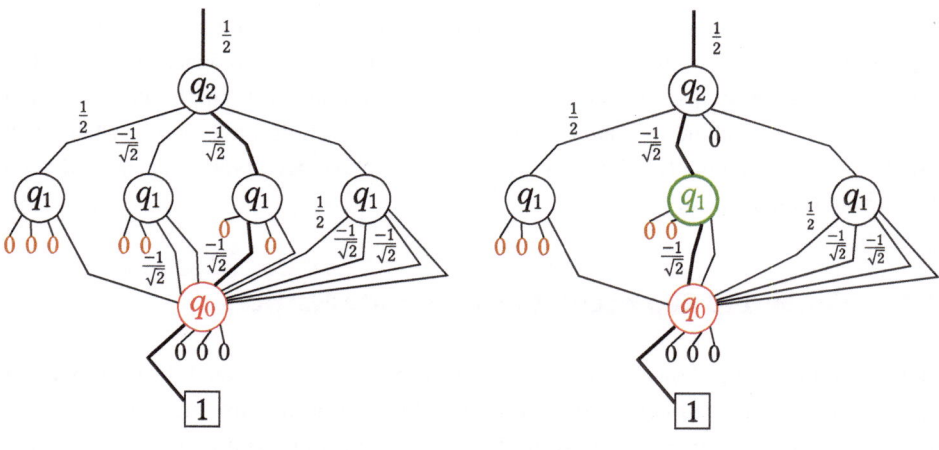

(a) Using established decomposition   (b) Using proposed decomposition

Fig. 4.3 Decision diagrams representing density matrices

the matrix (highlighted gray), the edge weight of the root edge ($\frac{1}{2}$) must be multiplied by the third edge of q_2 ($-\frac{1}{\sqrt{2}}$), the second edge of q_1 ($-\frac{1}{\sqrt{2}}$), and the first edge of q_0 (1) (see bold lines in figure), resulting in $\frac{1}{2} \cdot \frac{-1}{\sqrt{2}} \cdot \frac{-1}{\sqrt{2}} \cdot 1 = \frac{1}{4}$.

Knowing about the Hermitian property of density matrices, this representation can be optimized: Depending on the current position in the decomposition process, the upper right and lower left submatrices must be the complex conjugate transposes of each other. In these cases, the first successor still represents the upper left submatrix, while the third successor represents the lower right submatrix. Eventually, the second successor represents both the upper right and lower left submatrix in a shared fashion. More precisely, without loss of generality, the upper right submatrix is directly represented by the second successor, while the lower left submatrix is discarded and replaced by a "pointer" to the second node (a pointer that additionally stores that the corresponding successor node represents the complex conjugate transpose). Of course, the scheme is repeated recursively for all qubits again. Values of the respective matrix entries are then obtained again by multiplying the edge weights along the corresponding path—but dynamically modifying values when a pointer is taken, which indicates a complex conjugate transpose.

Example 4.6 Figure 4.3b shows the decision diagram representation of $\rho^\star$ that results when following the proposed decomposition scheme. As can be seen, the additional redundancies discussed before in Example 4.4 and framed in green in Fig. 4.2 can be exploited (in addition to the redundancies framed in orange and red that are already detected with the currently established scheme). To obtain a value of a matrix entry from this decision diagram, again, the edge weights of the corresponding path must be multiplied; but now taking a complex conjugate transpose into account when indicated by the respective pointer, i.e., by flipping the second and third edges and conjugating the edge weights. For example, to reconstruct the values $\frac{1}{4}$ from the lower left part of the matrix (highlighted gray), the edge weight of the root edge ($\frac{1}{2}$) must be multiplied by the right edge of q_2 ($-\frac{1}{\sqrt{2}}$), the second edge of q_1 ($-\frac{1}{\sqrt{2}}$), and the first edge of q_0 (1) (see bold lines in figure), resulting in $\frac{1}{2} \cdot (\frac{-1}{\sqrt{2}}) \cdot (\frac{-1}{\sqrt{2}}) \cdot 1 = \frac{1}{4}$. Notice that the path at q_1 is flipped—corresponding to a transpose of q_1 (so that the correct matrix element could be restored). Overall, this yields a decision diagram which requires one node less than the one discussed in Example 4.5 and shown in Fig. 4.3a—a reduction by approx. 17%.[1]

4.2.2 Applying Operations

Having an (optimized) quantum state representation of a density matrix obviously is good, but only really helps in quantum circuit simulation if operations can be applied to it, i.e., matrix-matrix-multiplication (see Definition 4.2). Fortunately, this can be efficiently realized using the proposed decision diagram scheme, as illustrated using matrix notation. The matrix-matrix-multiplication of the two quantum operations U and V can be written to correspond to

$$U \cdot M = \begin{bmatrix} U_0 & U_1 \\ U_2 & U_3 \end{bmatrix} \cdot \begin{bmatrix} M_0 & M_1 \\ M_2 & M_3 \end{bmatrix} \tag{4.10}$$

$$= \begin{bmatrix} U_0 \cdot M_0 & U_0 \cdot M_1 \\ U_2 \cdot M_0 & U_2 \cdot M_1 \end{bmatrix} + \begin{bmatrix} U_1 \cdot M_2 & U_1 \cdot M_3 \\ U_3 \cdot M_2 & U_3 \cdot M_3 \end{bmatrix}, \tag{4.11}$$

with the subscripted elements representing submatrices corresponding to the decomposition scheme presented in Example 4.6, i.e., the upper left submatrix corresponds to U_0, the upper right submatrix corresponds to U_1, the lower left submatrix corresponds to U_2, and the lower right submatrix corresponds to U_3 for U (the same applies to M). The necessary subproducts are determined by recursively continuing this process until submatrices of size one, i.e., complex numbers, remain onto which those operations can be applied directly. Hence, multiplication can be decomposed in a manner similar to the matrices yielding the (proposed) decision diagrams.

Since the off-diagonal elements of the density matrix are encoded into the same decision diagram substructure (although they are only the complex conjugate transpose of each other), some care has to be taken during the traversal of the decision diagram. Depending on the path taken in the decision diagram, the pointers have to be modified corresponding to a conjugate

---

[1] Note that 17% might not seem much. But considering that the example is rather small, it is significant. Discussions and evaluations, summarized and confirmed below, show that improvements of up to 50% are possible when exploiting the proposed potential.

complex transpose, i.e., the second and third edges have to be flipped and the edge weight must be conjugated. Besides that, however, operations can be applied in a similar fashion as before.

4.3   Resulting Potential

As shown above, exploiting the Hermitian characteristics as proposed in this chapter may lead to the detection of redundancies that were not caught before—yielding optimized, i.e., more compact representations for density matrices. However, the magnitude of the resulting potential strongly depends on the considered quantum states. This section briefly discusses the resulting potential in a conceptual fashion (before the derived conclusions are also confirmed empirically in the next section). To this end, two cases can be roughly distinguished:

(1) Cases with fewer redundancies yet: If density matrices are considered where current (established) decomposition schemes could not identify any (or not much) redundancies, substantial improvements are possible with the proposed scheme. In fact, even if nothing above the matrix diagonal is redundant (and, hence, offers potential for a compact representation), at least everything below the diagonal is definitely redundant and can be discarded. For a density matrix representing n qubits, this means that the representation of up to

$$\frac{(2^n - 1)2^n}{2} \tag{4.12}$$

entries of the matrix are redundant and can be avoided—yielding a reduction of up to 50% (but never 50% itself or more as the diagonal entries themselves still need to be represented). That is, the proposed scheme offers a particular improvement for instances where not much redundancy could be exploited yet.

(2) Cases which already include many redundancies: In contrast, if current (established) decomposition schemes already were able to detect many redundancies, the improvements achieved with the scheme proposed in this chapter are less or even nonexistent. In fact, if, e.g., the entries below the matrix diagonal are already available in a very compact fashion, exploiting the Hermitian characteristic hardly provides any further room for compaction. After all, there is a limit on how compact states can be represented.

That is, the proposed scheme offers substantially less or even no improvement for instances where a substantial amount of redundancy could already be exploited.

Example 4.7 The effects described above can even be seen in the example considered above for the state $\rho^\star$ and its density matrix representation in Fig. 4.2 as well as decision diagram representations in Fig. 4.3. In the first decomposition step (i.e., for q_1), the proposed scheme can exploit more redundancies, since the submatrices framed in green only need to be represented once (leading to one node less in the decision diagram). At the same time, in the subsequent decomposition steps, all redundancies identified by the proposed scheme can also be identified by the currently established scheme. That is, since they already allow for a compact representation, no further optimizations are obtained here. Overall, this does not lead to the best possible improvement of up to 50%, but still a significant one of approx. 17% which could not be exploited before.

4.4   Empirical Results

In order to also empirically evaluate the potential discussed above, the proposed scheme is implemented in C++ using the open-source decision diagram package from [119]. Afterwards, the performance of the current (established) decision diagram is compared with the proposed scheme when applied to quantum circuit simulation. In particular, the runtime of simulating different quantum circuits is measured, as well as the size of the resulting decision diagrams (i.e., their number of nodes). The experiments were executed on a system with 96 cores running at a clock frequency of 2.2 GHz and 1.5 TB of RAM.

As benchmarks, the *Quantum Fourier Transform* (QFT) [82] with an increasing number of qubits is used. The QFT is a common use-case and an essential part of several important quantum algorithms (e.g., Shor's factorization algorithm [96] or quantum phase estimation [82]), as well as Google's supremacy circuits, which are designed in such a way as to contain as few redundancies as possible [15]. In addition to that, a selection of further quantum circuits taken from the benchmark set of [71] is also used.

In Table 4.1, the results of the evaluation are summarized. For each considered quantum circuit, the number of qubits (#Q), the number of gates (#G), the node count of the final state is listed and the runtime of the quantum circuit simulation—for the established scheme (*Est.*) and the proposed scheme (*Prop.*). Finally, also the obtained improvements (*Improv.*) are provided.

The results clearly confirm the discussions of Sect. 4.3. If the current (established) decomposition scheme already identifies a large number of redundancies, the scheme proposed in this chapter does not provide any further optimization. This is particularity the case for the QFT circuits—here the involved decision diagrams mostly stay compact, so that no improvement can be reported. In contrast, if current (established) schemes were unable to detect many redundancies, substantial improvements of up to 49% can be achieved. This is particularly the case for Google's quantum supremacy circuits, which are designed in such a way as to contain as few redundancies as possible.

The benefits yielded by the proposed scheme for quantum circuit simulation increase with the absolute size of the involved decision diagrams. So, while no runtime improvement can be reported when the simulation is already very fast, considerable improvements of up to 53% are achieved for larger simulations.

Table 4.1 Empirical results

Circuit	#Q	#G	Nodes			Runtime		
			Est.	Prop.	Imp. (%)	Est.	Prop.	Imp. (%)
QFT	15	548	31	31	0	0.17	0.17	0
QFT	16	624	33	33	0	0.17	0.17	0
QFT	17	704	34	34	0	0.17	0.17	0
QFT	18	793	36	36	0	0.17	0.17	0
QFT	19	886	38	38	0	0.17	0.17	0
QFT	20	977	40	40	0	0.17	0.17	0
vqe_uccsd	4	220	170	100	41	0.17	0.17	0
vqe_uccsd	6	2282	2730	1428	47	2.37	1.9	19
vqe_uccsd	8	10808	43690	22100	49	172.28	117.92	31
qpe	9	150	2736	1434	47	0.17	0.17	0
qaoa	6	270	2730	1428	47	0.52	0.52	0
Quant. Sup.	7	368	10922	5588	48	5.58	3.16	43
Quant. Sup.	8	445	43690	22100	49	44.76	21.01	53
Quant. Sup.	9	442	174762	87892	49	292.71	177.18	39

4.5   Conclusion

Density matrices are an essential tool for noise-aware quantum circuit simulation, as they can represent noisy (i.e., mixed) quantum states. They are, therefore, essential in the forthcoming investigation on decision diagram-based noise-aware quantum circuit simulation. However, while current established quantum decision diagram schemes can directly represent density matrices, the Hermitian property of density matrices has not been exploited, yet. This was changed in this chapter.

To this end, an optimized decision diagram representation of density matrices that exploits the Hermitian property was proposed and discussed. The proposed optimization was implemented on top of a decision diagram-based quantum circuit simulator and empirically evaluated against a decision diagram-based quantum circuit simulator which does not utilize these improvements. The evaluation confirmed the previous discussion and showed that the improved density matrix representation is never larger than the established decision diagram matrix representation and up to 50% smaller. Furthermore, the more compact density matrix representation results in runtime improvements for quantum circuit simulation of up to 53%.

Overall, this chapter shows the potential of decision diagrams for noise-aware quantum circuit simulation and also sets the foundation for the investigation of decision diagram-based noise-aware quantum circuit simulation which is presented in the next part.

Part III
Approaches

Deterministic Simulation of Noise

The previous part established the basis for the investigation of noise-aware quantum circuit simulation by reviewing the current state-of-the-art on decision diagram-based quantum circuit simulation and by proposing a new decision diagram structure to represent density matrices. Building on this foundation, this part investigates the potential of decision diagrams for two distinct (but complementary) noise-aware quantum circuit simulation approaches.

The investigation is started in this chapter by considering noise effects in a deterministic manner using density matrices and Kraus operators. This formalism allows complete and deterministic consideration of errors during noise-aware quantum circuit simulation, but this comes with substantial additional complexity on top of the already exponentially hard problem of quantum circuit simulation. Thus, although there are quantum circuit simulators that implement this approach, their scalability is severely limited (e.g., [7, 25, 64, 89, 109, 111]). As shown in the previous part, decision diagrams can often keep the resource requirements of quantum simulation in check by exploiting redundancies within the classical descriptions of the corresponding quantum objects, which makes them a promising candidate for noise-aware quantum circuit simulation using density matrices and Kraus operators. However, thus far, their potential for this style of quantum circuit simulation has not been explored, yet.

In this chapter (based on [48, 50]), this is changed. To this end, the approach is presented first and the effects on quantum circuit simulation are discussed. Using the proposed decomposition scheme presented in Chap. 4, decision diagrams might remain compact in many cases—even if noise effects are considered. However, having a compact representation is not sufficient: Efficiently realizing the operations that describe the errors poses a substantial challenge. In order to mitigate those negative effects, an advanced approach for simulation is eventually proposed that can also handle noise effects efficiently. Based on that,

T. Grurl et al., *Noise-Aware Quantum Circuit Simulation with Decision Diagrams*, Synthesis Lectures on Engineering, Science, and Technology, https://doi.org/10.1007/978-3-031-71036-0_5

a decision diagram-based quantum circuit simulator is implemented and evaluated against a straightforward decision diagram-based implementation (without the proposed optimizations) and state-of-the-art solutions from IBM and Atos. The evaluation shows a mixed picture: The proposed optimization improves the performance compared to the straightforward implementation—often in the order of magnitudes. However, when comparing the proposed solution with the industry-grade tools, it is only considerably faster for one of the considered quantum applications, while it is slower for most others.

The remainder of the chapter is structured as follows: Sect. 5.1 illustrates the need for noise-aware quantum circuit simulation and introduces the approach for (deterministic) consideration of noise. Section 5.2 discusses how considering errors changes the way simulation is conducted and analyzes the resulting challenges for approaches based on decision diagrams. Based on these insights, an advanced simulation scheme is proposed in Sect. 5.3. The proposed noise-aware simulator is then thoroughly evaluated in Sect. 5.4 and the chapter is concluded in Sect. 5.5.

5.1   Deterministic Consideration of Noise

This chapter explores the potential of decision diagrams for deterministic noise-aware quantum circuit simulation. To this end, this section first shows how noise effects on current quantum computers distort the results of quantum algorithms—thereby motivating the need for noise-aware quantum circuit simulation—and then reviews how errors in quantum computing can be simulated using density matrices and Kraus operators.

5.1.1   Noise in Quantum Computers

The formalism introduced in Sect. 2.1 allows one to simulate the execution of *perfect* quantum hardware. However, since real quantum hardware is plagued by unavoidable errors, such a perfect simulation does not provide much insight into how a quantum circuit would behave when executed on a real quantum computer, as illustrated in the following example:

Example 5.1 In order to showcase how the simulation of a quantum circuit without consideration of errors only provides an incomplete picture, the circuit depicted in Fig. 5.1a is executed on real quantum hardware and the results are compared with the (simplified) simulation. More precisely, the circuit is executed using one of IBM's quantum computers [63] (that is, *ibmq_quito*, which is one of the IBM Falcon r4T (Version 1.1.41) processors) and simulated using Qiskit [89], without any consideration to noise effects. For both runs 1,024 samples were generated, and the respective measurement frequencies are plotted in Fig. 5.1b.

The measurements of the *perfect* simulation conforms almost perfectly with the ideal distribution (as calculated in Example 2.2), with the probability of measuring $|01\rangle$ and $|10\rangle$

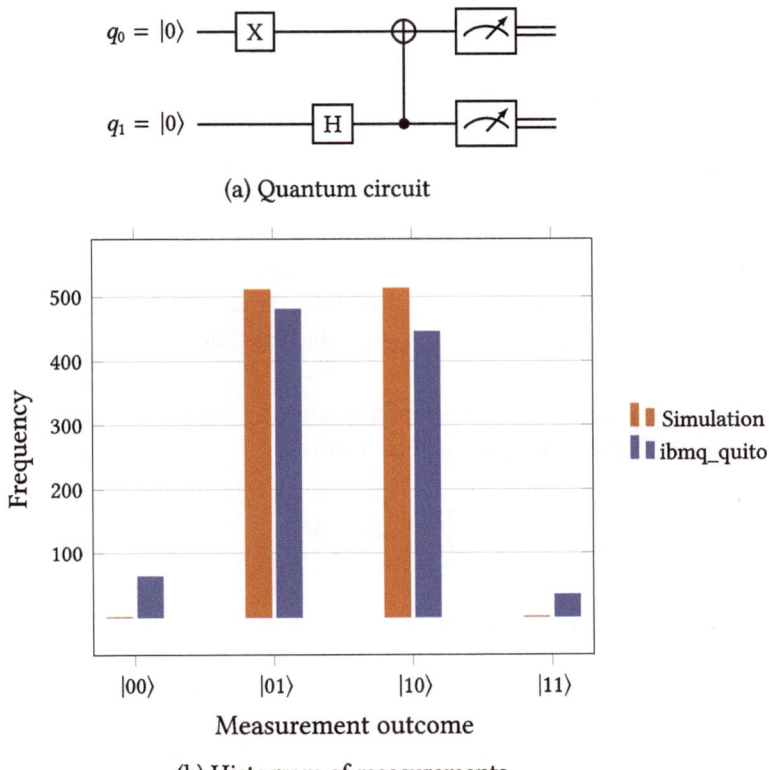

(a) Quantum circuit

(b) Histogram of measurements

Fig. 5.1 Noise effects on real quantum hardware

being both 50%. In contrast, the measurement distribution of the real quantum computer is much more noisy, and there is even a probability to measure $|00\rangle$ and $|11\rangle$, both of which are not even part of the ideal distribution.

The example shows how errors in today's quantum computers cause noise effects even for very small quantum circuits. For large quantum circuits—which are necessary to realize interesting functionality—these noise effects are even more pronounced. Thus, considering these noise effects during the development of quantum circuits is essential in order to gain insights into how they would behave when executed on real quantum hardware. The necessary formalism for such a noise-aware quantum circuit simulation is introduced next.

5.1.2 Representation of Errors

Considering errors in quantum circuit simulation requires

- a structure to handle noisy quantum states (i.e., state mixtures)
- and some means to represent and apply errors.

Regarding the first issue, density matrices (as introduced in Sect. 4.1.1) can be used as they provide a unified structure to represent arbitrary large state mixtures. Errors can be characterized using Kraus operators, which are defined as follows [82].

Definition 5.1 Using the operator-sum representation, an error is represented by a tuple $(E_0, E_1, \ldots, E_m)$ of operation elements (i.e., matrices) that satisfies the condition.

$$\sum_{i=0}^{m} E_i^\dagger E_i = \mathrm{I}. \tag{5.1}$$

Using this representation, errors in quantum computers can be characterized, such as [82].

Example 5.2 The depolarization error is given by $D = (E_0, E_1, E_2, E_3)$ with

$$E_0 := \sqrt{1 - \frac{3p}{4}} \cdot \begin{bmatrix} 1 & 0 \\ 0 & 1 \end{bmatrix}, E_1 := \sqrt{\frac{p}{4}} \cdot \begin{bmatrix} 0 & 1 \\ 1 & 0 \end{bmatrix}, \tag{5.2}$$

$$E_2 := \sqrt{\frac{p}{4}} \cdot \begin{bmatrix} 0 & -i \\ i & 0 \end{bmatrix}, E_3 := \sqrt{\frac{p}{4}} \cdot \begin{bmatrix} 1 & 0 \\ 0 & -1 \end{bmatrix}. \tag{5.3}$$

Amplitude damping, also known as the T1 decoherence error, can be described by $T1 = (E_0, E_1)$ with

$$E_0 := \begin{bmatrix} 1 & 0 \\ 0 & \sqrt{1-p} \end{bmatrix}, E_1 := \begin{bmatrix} 0 & \sqrt{p} \\ 0 & 0 \end{bmatrix} \tag{5.4}$$

and, finally, the T2 decoherence error can be described as $T2 = (E_0, E_1)$ with

$$E_0 := \sqrt{p} \cdot \begin{bmatrix} 1 & 0 \\ 0 & 1 \end{bmatrix}, E_1 := \sqrt{1-p} \cdot \begin{bmatrix} 1 & 0 \\ 0 & -1 \end{bmatrix}. \tag{5.5}$$

The variable p, which occurs in all operators, represents the probability that an error occurs [82]. This probability is a parameter of the specific quantum computer and has to be provided to the simulator.

Having such a representation of state mixtures by the means of density matrices and errors by the means of Kraus operators, errors can be applied to quantum states as follows:

Definition 5.2 Applying an error specified by the operators $(E_0, E_1, \ldots, E_m)$ to a quantum system given by the density matrix ρ yields the density matrix

$$\rho' = \sum_{i=0}^{m} E_i \rho E_i^\dagger. \tag{5.6}$$

Example 5.3 Consider again the quantum register

$$|\varphi'\rangle = \begin{bmatrix} 0 & \frac{1}{\sqrt{2}} & \frac{1}{\sqrt{2}} & 0 \end{bmatrix}^\top \tag{5.7}$$

from Example 2.2. Using Definition 4.1, the density matrix $\rho = |\varphi'\rangle\langle\varphi'|$ is given by

$$\underbrace{\begin{bmatrix} 0 \\ \frac{1}{\sqrt{2}} \\ \frac{1}{\sqrt{2}} \\ 0 \end{bmatrix}}_{|\varphi'\rangle} \cdot \underbrace{\begin{bmatrix} 0 & \frac{1}{\sqrt{2}} & \frac{1}{\sqrt{2}} & 0 \end{bmatrix}}_{\langle\varphi'|} = \underbrace{\begin{bmatrix} 0 & 0 & 0 & 0 \\ 0 & \frac{1}{2} & \frac{1}{2} & 0 \\ 0 & \frac{1}{2} & \frac{1}{2} & 0 \\ 0 & 0 & 0 & 0 \end{bmatrix}}_{\rho}. \tag{5.8}$$

Analogous to Example 4.1, an amplitude damping (T1) error is simulated that affects q_0 of the state ρ with a probability of 2% ($p = 0.02$), but now, Kraus operators are used to simulate the effect. Using the representation of the T1 error provided in Eq. (5.4), the error is applied to q_0 of ρ like so:

$$\underbrace{\begin{bmatrix} 1 & 0 & 0 & 0 \\ 0 & 0.989949 & 0 & 0 \\ 0 & 0 & 1 & 0 \\ 0 & 0 & 0 & 0.989949 \end{bmatrix}}_{(I \otimes E_0)} \cdot \underbrace{\begin{bmatrix} 0 & 0 & 0 & 0 \\ 0 & \frac{1}{2} & \frac{1}{2} & 0 \\ 0 & \frac{1}{2} & \frac{1}{2} & 0 \\ 0 & 0 & 0 & 0 \end{bmatrix}}_{\rho} \cdot \underbrace{\begin{bmatrix} 1 & 0 & 0 & 0 \\ 0 & 0.989949 & 0 & 0 \\ 0 & 0 & 1 & 0 \\ 0 & 0 & 0 & 0.989949 \end{bmatrix}}_{(I \otimes E_0)^\dagger} + \tag{5.9}$$

$$\underbrace{\begin{bmatrix} 0 & 0.141421 & 0 & 0 \\ 0 & 0 & 0 & 0 \\ 0 & 0 & 0 & 0.141421 \\ 0 & 0 & 0 & 0 \end{bmatrix}}_{(I \otimes E_1)} \cdot \underbrace{\begin{bmatrix} 0 & 0 & 0 & 0 \\ 0 & \frac{1}{2} & \frac{1}{2} & 0 \\ 0 & \frac{1}{2} & \frac{1}{2} & 0 \\ 0 & 0 & 0 & 0 \end{bmatrix}}_{\rho} \cdot \underbrace{\begin{bmatrix} 0 & 0 & 0 & 0 \\ 0.141421 & 0 & 0 & 0 \\ 0 & 0 & 0 & 0 \\ 0 & 0 & 0.141421 & 0 \end{bmatrix}}_{(I \otimes E_1)^\dagger} = \tag{5.10}$$

$$
\underbrace{\begin{bmatrix} 0 & 0 & 0 & 0 \\ 0 & 0.49 & 0.494975 & 0 \\ 0 & 0.494975 & 0.5 & 0 \\ 0 & 0 & 0 & 0 \end{bmatrix}}_{(I \otimes E_0) \cdot \rho \cdot (I \otimes E_0)^{\dagger}} + \underbrace{\begin{bmatrix} 0.01 & 0 & 0 & 0 \\ 0 & 0 & 0 & 0 \\ 0 & 0 & 0 & 0 \\ 0 & 0 & 0 & 0 \end{bmatrix}}_{(I \otimes E_1) \cdot \rho \cdot (I \otimes E_1)^{\dagger}} = \tag{5.11}
$$

$$
\underbrace{\begin{bmatrix} 0.01 & 0 & 0 & 0 \\ 0 & 0.49 & 0.494975 & 0 \\ 0 & 0.494975 & 0.5 & 0 \\ 0 & 0 & 0 & 0 \end{bmatrix}}_{\rho'} \tag{5.12}
$$

The resulting density matrix ρ' contains the effect of the error: While the probability of measuring $|10\rangle$ is still 0.5 , the probability of measuring $|01\rangle$ has dropped to 0.49 and now there is a probability of 0.01 to measure $|00\rangle$. Therefore, the probability that q_0 is measured $|0\rangle$ has increased by 2%, reflecting the amplitude damping error assumed above.

Having this mathematical description of noise, a wide range of errors affecting present-day quantum computers can be considered (for each error, just a corresponding representation in terms of Kraus operators needs to be provided). Thus, the formalism can be used as a basis for mimicking the execution of real quantum computers for quantum circuit simulation.

5.2   Effects on the Simulation

Using density matrices and Kraus operators, approaches for quantum circuit simulation can be accordingly extended. For straightforward approaches as reviewed in Sect. 3.1.1 (representing the corresponding vectors and matrices in terms of one- and two-dimensional arrays, respectively) this is particularly straightforward and revolves mainly around changing the state representation from vectors to matrices. This can be addressed by extending the underlying data structure (i.e., arrays) accordingly. The newly required operations, i.e., matrix-matrix multiplications and the addition of matrices, can be directly implemented and are often already supported by the underlying libraries anyway. Thus, many such quantum circuit simulators supporting the consideration of noise are already available (see, e.g., [7, 25, 64, 89, 109, 111]). However, switching from state vectors to density matrices makes the curse of dimensionality and, by this, the resulting complexity even worse—severely limiting the corresponding approaches.

The question remains as to how the quantum circuit simulation based on decision diagrams is affected by extending the formalism. Since compactness constitutes one of the main advantages of decision diagram-based quantum circuit simulation, an important question becomes how (much) this is compromised when working with density matrices.

The evaluation of the proposed decision diagram-based density matrix representation in the previous chapter (see Sect. 4.4), sheds some light on this and suggests that the advantage of decision diagrams is not completely lost when density matrices are used to represent the quantum state.

However, having a compact representation is not enough, as efficient realization of matrix-matrix operations (mainly multiplication and addition), as presented in Sect. 5.1.2, is also necessary. In the case of multiplication, the concepts proposed in Chap. 4 can be used. Matrix-matrix addition, however, which is required for applying errors as defined in Eq. (5.6), has thus far only been a suboperation of multiplication and turns out to be particularly challenging.

Recall that adding two matrices is done by adding all elements that share the same index. Thus, access to *all* matrix elements is required. As illustrated in Sect. 3.2.2, this constitutes a bottleneck when working with quantum decision diagrams, since each matrix element is encoded into the tree structure and must be restored for the operation. Accessing *all* matrix elements in order to add them requires decompression and, therefore, traversal of the entire decision diagram. This is in contrast to multiplication, where, depending on the applied quantum operation, multiplication can be applied in some cases without decompressing any element at all (as illustrated in Sect. 3.2.1).

Overall, the investigation indicates that the decision diagram structure may also be suitable when used for noise-aware quantum circuit simulation using density matrices. However, due to the extended formalism, new challenges arise which must be addressed, namely, how to efficiently apply errors to the state without having to traverse the entire decision diagram.

5.3   Advanced Simulation Approach

A major challenge when considering noise effects during decision diagram-based quantum circuit simulation is to efficiently realize the necessary operations (particularly Eq. (5.6)). Applying errors revolves around adding matrices, which requires access to *all* matrix elements and thus triggers the complete traversal of the involved decision diagrams. This causes an exponential overhead during the simulation, which severely impacts the performance of corresponding approaches. To address this challenge, alternatives are investigated that either completely avoid adding matrices or, at least, involve only (smaller) submatrices. The investigations finally lead to an alternative whose main idea is based on the following three observations:

First, adding matrices to apply errors can, in some cases, be avoided altogether. Specifically, the T2 error, which is characterized by the Kraus operators in Eq. (5.5), can be realized by multiplications only, i.e., the effect of this error on a single qubit can be captured by

$$\begin{bmatrix} a & b \\ c & d \end{bmatrix} \longmapsto \begin{bmatrix} a & (2p-1)b \\ (2p-1)c & d \end{bmatrix}, \tag{5.13}$$

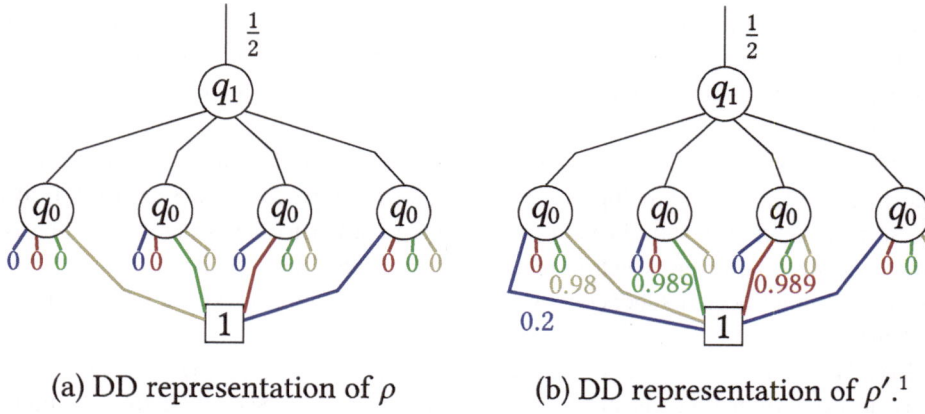

(a) DD representation of ρ (b) DD representation of ρ'.[1]

Fig. 5.2 Decision diagram (DD) representation of state vectors and density matrices

where p represents the probability that an error occurs. The colors in Eq. (5.13) illustrate how the matrix elements relate to decision diagram edges of Fig. 5.2a. That is, the matrix elements a, b, c, and d relate to the four outgoing edges of a decision diagram node from left to right. Therefore, applying the T2 error can be realized by applying a factor to specific edges of the decision diagram, for which *no* amplitude has to be decoded at all.[1]

Second, sequentially applying errors, as defined in Eq. (5.6), requires accessing the decision diagram multiple times. This process can be substantially improved by explicitly enforcing the error effects directly on the corresponding nodes of the affected qubits. In doing so, all desired effects can be applied to all qubits with just one traversal of the decision diagram.

Third, because of the tree-like structure of decision diagram representations of quantum states, applying operations to the quantum state often only affects parts of the decision diagram. More precisely, in a decision diagram, every qubit is represented by one or more nodes and applying an operation to a qubit only modifies the outgoing edges from the nodes representing this specific qubit. Predecessor nodes are only indirectly affected during the normalization process.

Based on the observations above, an advanced scheme is proposed to realize Eq. (5.6) for applying errors to quantum decision diagrams. The key idea is to directly modify the decision diagram to reduce overhead and merge separate operations. More precisely, it follows from the third observation that applying operations to qubits affects only the outgoing edges of nodes representing this qubit. Using this knowledge, it becomes straightforward to predict how each operation will modify the decision diagram. Consequently, this allows one to aggregate distinct operations—be it matrix-matrix-multiplications/additions, applying separate noise effects, or applying noise operations to different qubits—into a single operation. In addition, instead of conducting these operations on fully sized decision dia-

---

[1] The values 0.989 are rounded from 0.98995 in the diagram due to limited space.

grams, by directly applying the operations to the respective nodes, it suffices to only modify sub-decision diagrams, which may be considerably smaller.

Example 5.4 To illustrate this idea, consider again Example 5.3 where an amplitude damping (T1) error with a probability of 2% ($p = 0.02$) is applied to qubit q_0 of the quantum state ρ. However, now the error is directly applied to the decision diagram representation of ρ which is provided in Fig. 5.2a. The effects of the T1 error with $p = 0.02$ are given by the following transformation

$$\begin{bmatrix} a & b \\ c & d \end{bmatrix} \longmapsto \begin{bmatrix} a + 0.02 \cdot d & \sqrt{0.98} \cdot b \\ \sqrt{0.98} \cdot c & 0.98 \cdot d \end{bmatrix}. \tag{5.14}$$

The color coding once again illustrates how each matrix element relates to one of the outgoing edges of a decision diagram node Fig. 5.2a, i.e., the matrix elements a, b, c, and d represent the four outgoing edges of a decision diagram node from left to right. Modifying all nodes labeled q_0 accordingly leads to the new decision diagram depicted in Fig. 5.2a. This new decision diagram represents ρ after the T1 error with $p = 0.02$ has been applied and is equal to ρ' from Example 5.3.

Overall, by extending the formalism to density matrices and state mixtures, deterministic consideration of errors is possible. While this comes with a substantially increased complexity of the state description, the main advantage of decision diagram-based state representation—compactness—is not completely lost. To address the additional complexity of applying error operations (compared to standard quantum operations), an alternative, more efficient scheme is proposed. In the next section, this scheme is evaluated and the proposed decision diagram-based approach is compared against state-of-the-art solutions.

5.4 Empirical Results

In order to empirically evaluate the proposed noise-aware quantum circuit simulation approaches, the concepts presented here are implemented in C++, on top of the open-source decision diagram package taken from [119, 121] and extensively evaluated. To this end, the setup of the evaluation is described next. Afterwards, the performance is empirically evaluated in two scenarios, i.e., first the proposed optimizations are evaluated and secondly the proposed solution is compared against state-of-the-art simulators from IBM and Atos.

5.4.1 Setup

To obtain meaningful results, a range of different benchmark sets is used. More specifically, the *Entanglement* circuit is used—a quantum algorithm that generates the GHZ state—as well as the *Quantum Fourier Transform* (QFT) [82], both with an increasing number of qubits. In addition to this, also the *QASMBench* benchmark suite is used in the experiments which contains a broad range of different quantum circuit algorithms [71]. To ensure a fair comparison, all quantum circuits are translated into QASM basic gates (see [26]) prior to the experiments.

In the experiments, the noise effects reviewed in Sect. 2.2 are assumed. That is, gate errors are approximated using depolarization with a probability of 0.1%, amplitude damping error (T1) with 0.2% probability, and phase-flip error (T2) with 0.1% probability. Each noise effect is (probabilistically) applied to the qubit, whenever it is used and the error probabilities are doubled for multi-qubit operations.

All experiments are conducted on a system with 96 cores running at a clock frequency of 2.2 GHz and 1.5 TB of RAM. As methods, the *LinAlg* simulator of Atos' *QLM* [7], the *density matrix* simulators of IBM's *Qiskit* [89], and the proposed decision diagram-based solution are considered. While the QLM simulator runs directly on the system, Qiskit and the proposed simulator are ported to the machine using Docker [75]. Docker is used since its virtualization overhead is negligible [36], which ensures a fair comparison. For all experiments, a timeout of half an hour (1800 s) is applied. In the following summary, only the most relevant benchmarks are listed, i.e., benchmarks that could be simulated by *all* considered approaches in less than a few seconds, and benchmarks that no simulator could simulate within the given time limit are omitted.

5.4.2 Comparison Between Implementations

First, the advantages of the proposed optimizations are evaluated. To this end, two versions of the decision diagram-based solution are implemented: One in which the necessary concepts are implemented in a basic fashion (i.e., as presented in Sect. 5.1.2) and one in which the optimizations presented in Sect. 5.3 are implemented.

In Table 5.1 the results are provided, i.e., the name of the benchmark, the number of qubits (#Q), the number of gates (#G), the runtime for the basic and the optimized implementation, as well as the relative runtime improvements with the optimization compared to the basic implementation.

The optimizations substantially improve the runtime for all circuits with more than a few seconds runtime with performance gains that are even in the order of *magnitudes* for some circuits.

5.4.3 Comparison to Related Work

In a second series of evaluations, the proposed simulation approach (in its optimized version) is compared to other quantum circuit simulators. Two simulators are used that constitute good

Table 5.1 Optimization for deterministic simulation

Benchmark	#Q	#G	Basic [s]	Optimized [s]	Runtime Imp.
qaoa	6	270	1.81	0.75	58.56%
vqe_uccsd	8	10808	>1800.00	1301.89	–
qpe	9	150	R130.25	6.24	95.20%
QFT	9	192	37.53	3.66	90.24%
QFT	10	241	935.09	73.23	92.16%
QFT	11	291	>1800.00	1382.62	–
multiply	13	124	>1800	303.94	83.11%
entanglement	27	27	1299.19	190.8	85.31%
entanglement	28	28	>1800.00	318.05	–
entanglement	29	29	>1800.00	460.37	–
entanglement	30	30	>1800.00	708.06	–

representatives of the current state of the art, i.e., the *LinAlg* simulator of Atos' *Quantum Learning Machine* (QLM) [7], as well as the *density matrix* simulators of IBM's *Qiskit* [89]. In contrast to the other evaluation—where decoherence and depolarization errors have been considered together—here additionally the scenario with depolarization errors only is considered. This allows to evaluate how the different noise models affect the performance.

In Table 5.2, the results of the experiments are provided.[2] For each experiment the name of the benchmark, the number of qubits (#Q), number of gates (#G), as well as the required runtime for simulating it with the QLM, Qiskit, and the decision diagram-based solution in seconds is provided. Furthermore, no results from Atos' QLM simulator are listed for the QASMBench benchmarks, due to compatibility issues.

The results show a mixed picture when comparing the proposed simulator against the industry-grade solutions. The decision diagram-based simulator is only considerably faster (and more scalable) for simulating the entanglement circuit. The improvements in this case show that the main advantage of decision diagram-based simulators—namely their compactness—is not completely lost when simulating with density matrices. On the flip side, the poor results for the remaining benchmarks show that the additional complexity, that is introduced by simulating noise effects, cannot yet completely be mitigated by the current type of decision diagrams. Here, the current state-of-the-art, i.e., Atos' QLM and IBM's Qiskit simulators, still seem to provide the better solutions (or, at least, solutions that are not that much affected by the considered benchmark and/or noise model). By this, these results show that, although decision diagrams may provide some promises for an efficient deterministic consideration (as seen by the entanglement benchmark), they still struggle

---

[2] When simulating larger instances of the QFT circuit with Qiskit the simulation terminated due to insufficient memory. This is denoted in the table as "memory".

Table 5.2 Comparison to state-of-the-art deterministic quantum circuit simulators

Noise model	Entanglement circuit				QFT circuit				QASMBench benchmarks			
	#Q/#G	QLM	Qiskit	Prop.	#Q/#G	QLM	Qiskit	Prop.	Benchmark	#Q/#G	Qiskit	Prop.
Decoherence and Depolarization	16/16	709.22	62.88	0.86	7/116	1.43	1.50	0.29	basis_trotter	4/1626	2.30	0.41
	17/17	>1800.00	257.29	1.79	8/153	1.55	1.50	0.76	qaoa	6/270	1.49	0.75
	18/18	–	1089.40	3.64	9/192	1.94	1.51	3.66	vqe_uccsd	6/2282	2.53	4.01
	19/19	–	memory	4.9	10/241	3.43	1.62	73.23	vqe_uccsd	8/10808	8.51	1301.89
	20/20	–	–	8.97	11/291	11.24	1.91	1382.62	ising	10/480	1.62	>1800.00
	...	–	–	...	12/346	46.57	2.88	>1800.00	sat	11/679	2.94	>1800.00
	29/29	–	–	460.37	13/410	219.82	7.18	–	seca	11/282	1.91	>1800.00
	30/30	–	–	708.06	14/475	1036.49	26.93	–	multiply	13/124	3.76	303.94
	31/31	–	–	953.95	15/548	>1800.00	111.18	–	bv	14/43	5.19	>1800.00
	32/32	–	–	1375.3	16/624	–	488.74	–	Multipler	15/574	208.28	>1800.00
	33/33	–	–	>1800.00	17/704	–	>1800.00	–	cc	18/73	1600.65	>1800.00
Depolarization	16/16	166.48	15.97	0.17	8/153	1.21	1.25	0.17	basis_trotter	4/1626	1.94	0.29
	17/17	703.10	63.22	0.17	7/192	1.33	1.26	0.17	qaoa	6/270	1.26	0.64
	18/18	>1800.00	256.06	0.29	9/241	1.71	1.26	0.40	vqe_uccsd	6/2282	2.29	6.81
	19/19	–	1085	0.4	10/291	3.07	1.37	1.11	vqe_uccsd	8/10808	8.6	>1800.00
	20/20	–	>1800.00	0.52	11/346	10.98	1.66	4.25	ising	10/480	1.39	>1800.00
	...	–	–	...	12/410	46.87	2.87	60.38	sat	11/679	2.58	>1800.00
	50/50	–	–	1237.37	13/475	219.30	7.1	185.68	seca	11/282	1.54	1421.27
	51/51	–	–	1354.58	14/548	1025.74	26.92	>1800.00	multiply	13/124	3.69	66.69
	52/52	–	–	1401.08	15/624	>1800.00	112.52	–	bv	14/43	5.10	31.01
	53/53	–	–	1390.98	16/704	–	493.35	–	multipler	15/574	208.7	>1800.00
	54/54	–	–	>1800.00	17/793	–	>1800.00	–	cc	18/73	1586.38	>1800.00

with providing a robust solution that works in general—motivating further research, e.g., towards alternative decision diagram types and/or further optimization dedicated to the compact representation of density matrices.

5.5 Conclusion

Building on the foundation established in the previous part, this chapter explored the potential of decision diagrams for deterministic noise-aware quantum circuit simulation using density matrices and Kraus operators.

To this end, it was first introduced how errors can be simulated using this scheme and it was investigated how the necessary functionality can be implemented on top of a decision diagram-based quantum circuit simulator. The investigation showed that the compact representation of quantum states is not necessarily harmed by density matrices, but other challenges were identified with respect to conducting the required operations in a runtime efficient fashion. In order to address these challenges, an advanced method was proposed to realize the necessary functionality, and the proposed simulator was then implemented in C++.

The implemented decision diagram-based deterministic simulator was empirically evaluated against a straightforward decision diagram-based implementation (without the proposed optimizations) and state-of-the-art solutions from IBM and Atos. The results confirmed the usefulness of the proposed optimization, which improves the performance considerably compared to the straightforward implementation—often in the order of magnitudes. However, when comparing the proposed solution against the industry-grade tools, it was only considerably faster for one of the considered quantum applications, while it was slower for most others.

Overall, the chapter showed the potential of decision diagrams for noise-aware quantum circuit simulation, thereby motivating further research in this area. The implementation used for the evaluation is available as open-source under the MIT license under https://github.com/cda-tum/mqt-ddsim.

Stochastic Simulation of Noise   6

The previous chapter explores the potential of decision diagrams for deterministic noise-aware quantum circuit simulation using density matrices and Kraus operators. The evaluation showed that using decision diagrams for this task results in exponential speed-ups for one of the considered applications, compared to state-of-the-art solutions. However, at the same time, the industry-grade solutions still seem to perform better for most applications. To address this shortcoming, this chapter investigates an alternative approach for noise-aware quantum circuit simulation that avoids the use of density matrices by considering the errors in a stochastic fashion.

To achieve this, it is assumed that errors occur randomly throughout individual simulation runs. Afterwards, the true effects of errors on the quantum computation are approximated by forming empirical averages over multiple simulation runs (Monte Carlo). This provides a conceptually suitable and mathematically rigorous solution for the classical simulation of noisy quantum computations, which can be directly implemented on top of existing simulators.

However, severe challenges remain. First and foremost, performing a single simulation run requires repeated multiplications of matrices and vectors, which are of exponential size with regard to the number of tracked qubits. To make matters worse, a single stochastic simulation run does not adequately capture error effects. Instead, a sufficiently large number of independent simulation runs must be conducted to form empirical averages that accurately reflect the true quantum evolution. Together, both factors limit existing solutions with respect to efficiency and scalability, e.g., [7, 25, 54, 89].

This chapter (based on [50, 52]), investigates the potential of decision diagrams for noise-aware stochastic quantum circuit simulation. To this end, the necessary formalism is first introduced, and the effects on the decision diagram-based quantum circuit

© The Author(s), under exclusive license to Springer Nature Switzerland AG 2025   65
T. Grurl et al., *Noise-Aware Quantum Circuit Simulation with Decision Diagrams*,
Synthesis Lectures on Engineering, Science, and Technology,
https://doi.org/10.1007/978-3-031-71036-0_6

simulation are investigated. In order to mitigate those negative effects, advanced schemes for decision diagram-based simulation are eventually proposed that can also handle noise effects efficiently. Based on that, a quantum circuit simulator is implemented and compared with a straightforward decision diagram-based implementation (without the proposed optimizations), state-of-the-art solutions from IBM and Atos, and the deterministic decision diagram-based simulator proposed in Chap. 5. The findings confirm the advantages of the proposed optimization, show considerable improvements (speed-ups of several magnitudes and much better scalability) compared to the industry-grade solutions for many circuits, and also reveal considerable speed-ups compared to the deterministic solution proposed in the previous chapter.

The remainder of this chapter is structured as follows: Sect. 6.1 discusses how errors can be simulated using a stochastic approach. Then, Sect. 6.2 discusses the challenges of the proposed scheme on quantum circuit simulation, and Sect. 6.3 proposes advanced schemes to address these challenges in a decision diagram-based stochastic simulator. The proposed simulator is then extensively evaluated in Sect. 6.4 and, finally, the chapter is concluded in Sect. 6.5.

6.1   Stochastic Consideration of Errors

This chapter explores the potential of decision diagrams for stochastic noise-aware quantum circuit simulation. To this end, this section introduces the idea and then presents the necessary formalism.

6.1.1   Stochastic Simulation of Errors

While density matrices and Kraus operators allow one to deterministically mimic errors of real quantum computers, this formalism also introduces considerable overhead into the already exponentially hard task of quantum circuit simulation. Fortunately, this hard task is often an overkill anyway, since real quantum computers do not operate deterministically, as illustrated in the following example.

Example 6.1 Consider the quantum circuit depicted in Fig. 6.1a.[1] When this circuit is executed on a real quantum computer, (unavoidable) errors deteriorate the result, as shown in Example 5.1. These hardware errors can be considered during quantum circuit simulation. For example, whenever a qubit is used in an operation, an amplitude damping error can be simulated, which affects the used qubit with 2% probability. Simulating this circuit with consideration to such an amplitude damping error using density matrices and Kraus operators

---

[1] All values in this example have been rounded to 4 decimal places for readability.

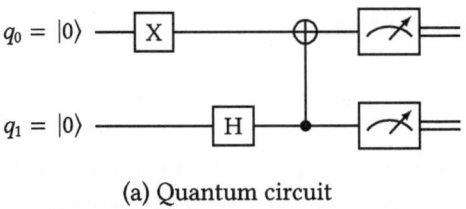

(a) Quantum circuit

$$\begin{bmatrix} 0.0298 & 0. & 0. & 0.0097 \\ 0. & 0.49 & 0.4754 & 0. \\ 0. & 0.4754 & 0.4708 & 0. \\ 0.0097 & 0. & 0. & 0.0094 \end{bmatrix}$$

(b) Density matrix

$\{(0.0198, |00\rangle), (0.0002, |10\rangle), (0.0098, |01\rangle),$

$(0.9508, 0.7107 \cdot |01\rangle + 0.7035 |10\rangle),$

$(0.0194, 0.7177 \cdot |01\rangle + 0.6963 |11\rangle)\}$

(c) State ensemble

Fig. 6.1 Result of noisy quantum circuit simulation (All values in this example have been rounded to 4 decimal places for readability)

(as presented in Chap. 5), results in the density matrix depicted in Fig. 6.1b. However, a real quantum computer would not calculate the full state mixture, but rather apply the quantum gates, and errors would randomly affect the state. Thus, the final state of such an execution can be viewed as one (possible) state of the final state mixture. For this example, the final state mixture can be described by the state ensemble in Fig. 6.1c. Having this state ensemble now allows one to fully characterize the final state without the need of using any density matrices.

Example 6.1 illustrates how noisy quantum states can be represented without density matrices. However, generating the final state ensemble for circuits consisting of more than a few qubits quickly becomes unfeasible. Fortunately, by sacrificing the deterministic description of the quantum state, calculating the full state ensemble can be avoided. The basic idea is as follows: Whenever an error could occur, a die is rolled, and the error is applied depending on the error probability and the die roll. After simulating in such a fashion, one possible final state is generated, and by repeating this simulation step more of these (possible) final states are created. Finally, having generated a sufficient number of (possible) final states, the real final state can be approximated using Monte Carlo sampling. This allows to keep the actual simulation runs more efficient, but may require a substantial number of runs for proper sampling and, after all, remains a stochastic (rather than deterministic) approach.

6.1.2 Approximation of Errors

Realizing a stochastic quantum circuit simulator requires

- (1) a formalism that allows one to impose errors on quantum state representations, and
- (2) a mathematical model to predict how many samples are necessary for a reliable result.

Addressing (1), noise effects occurring during quantum computing can be viewed as (unwanted) operations to the quantum state. Using this idea allows one to reuse the procedures for quantum circuit simulation introduced in Sect. 3.1. That is, quantum states and operations are represented by vectors and matrices. Operations are then applied to quantum states using matrix-vector-multiplications. By this, errors can be viewed as additional operations that only occur with some probability p. Such an erroneous operation leaves the state in the mixture $\{(p_i, |\psi_i\rangle)|1 \le i \le k\}$, with the system being in the state $|\psi_i\rangle$ with probability p_i (see Definition 2.5). From this state mixture, all but one state is discarded and the simulation is continued. The state mixture depends on the probability of the error, as well as the type of error.

Errors can be represented in terms of operations on the state: Gate errors can be mimicked by depolarization errors, which are characterized by randomly applying an X, Y, Z, or I operation to the state. Similarly, phase-flip (T2) decoherence errors can be realized by applying a Z operation. Finally, due to its irreversible nature, mimicking the effect of amplitude damping (T1) decoherence errors, cannot be handled so easily. The error is mimicked by applying either

$$\begin{bmatrix} 1 & 0 \\ 0 & \sqrt{1-p} \end{bmatrix} \text{ or } \begin{bmatrix} 0 & \sqrt{p} \\ 0 & 0 \end{bmatrix} \tag{6.1}$$

to the state. Afterwards, the resulting quantum state must be normalized again.

Example 6.2 To illustrate how different kinds of errors can be captured, a depolarization, amplitude damping (T1), and phase-flip (T2) error is applied to a quantum state. More precisely, each error is applied to q_0 of the two-qubit register

$$|\varphi'\rangle = \begin{bmatrix} 0 & \frac{1}{\sqrt{2}} & \frac{1}{\sqrt{2}} & 0 \end{bmatrix}^{\top} = \frac{1}{\sqrt{2}} \cdot (|01\rangle + |10\rangle) \tag{6.2}$$

from Example 2.2. with 2% probability ($p = 0.02$). The resulting state mixture after applying the depolarization error to $|\varphi'\rangle$ is

$$\{(0.98, \tfrac{1}{\sqrt{2}} \cdot (|01\rangle + |10\rangle), \tag{6.3}$$

$$(0.005, \tfrac{1}{\sqrt{2}} \cdot (|01\rangle + |10\rangle), \tag{6.4}$$

$$(0.005, \tfrac{1}{\sqrt{2}} \cdot (|00\rangle + |11\rangle), \tag{6.5}$$

$$(0.005, \tfrac{-i}{\sqrt{2}} \cdot (|00\rangle - |11\rangle), \tag{6.6}$$

$$(0.005, \tfrac{-1}{\sqrt{2}} \cdot (|01\rangle - |10\rangle)\}, \tag{6.7}$$

the state mixture after applying an amplitude damping (T1) error to $|\varphi'\rangle$ is

$$\{(0.01, |00\rangle), (0.99, 0.703526 \cdot |01\rangle + 0.710669 \cdot |10\rangle\}, \tag{6.8}$$

and, finally, the state mixture after applying an phase-flip (T2) error to $|\varphi'\rangle$ is

$$\{(0.98, \tfrac{1}{\sqrt{2}} \cdot (|01\rangle + |10\rangle), (0.02, \tfrac{-1}{\sqrt{2}} \cdot (|01\rangle - |10\rangle)\}. \tag{6.9}$$

After calculating the respective state mixture, the next steps are the same for all types of errors. That is, one of the states is randomly chosen (with respect to its probability), and all other states are discarded. Then the simulation is continued only with the chosen state.

Having a scheme to stochastically apply errors, the next question becomes (2), i.e., how many samples must be generated to get a good prediction of the true probabilities. Recall that by simulating in a stochastic fashion, *one possible* final state $|\tilde{\psi}\rangle$ is calculated from the actual mixture $\{(p_i, |\psi_i\rangle)\}$: $|\tilde{\psi}\rangle = |\psi_i\rangle$ with probability p_i. The true distribution of sampled output states can be approximated by forming empirical averages using Monte Carlo approximation. This scheme is especially well suited to accurately learn properties of the final state (distribution) without the need to keep track of the complete (exponential) distribution. In quantum computing, interesting properties (e.g., the outcome probability of measurements) can often be described as a quadratic function in the state vector, more precisely $o_l = |\langle \omega_l | \psi \rangle|^2$, with ω_l being for example the vector representing a basis state. In the case of a probabilistic state mixture $\{(p_i, |\psi_i\rangle)\}$, such a quadratic property becomes

$$o_l = \sum_i p_i |\langle \omega_l | \psi_i \rangle|^2 \tag{6.10}$$

and can be approximated by an empirical average $\hat{o}_l$ over M samples $|\tilde{\psi}_j\rangle$ from the distribution, i.e.,

$$\hat{o}_l = \tfrac{1}{M} \sum_{j=1}^{M} \left| \langle \omega_l | \tilde{\psi}_j \rangle \right|^2 \quad \text{(Monte Carlo)}. \tag{6.11}$$

Furthermore, the same sample collection $\left\{ |\tilde{\psi}_1\rangle, \ldots, |\tilde{\psi}_M\rangle \right\}$ can be used to estimate multiple quadratic properties at the same time.

Using that, a sufficient sample set is motivated by the following theorem (taken from [52]).

Theorem 6.1 *Fix a collection of L (arbitrary) quadratic properties (6.10), as well as $\epsilon \in (0, 1)$ and $\delta \in (0, 1)$. Then,*

$$M = \frac{\ln\left(\frac{2L}{\delta}\right)}{2\epsilon^2} \tag{6.12}$$

state samples suffice to accurately approximate all *target properties with high confidence, i.e.,* $\max_l |\hat{o}_l - o_l| \leq \epsilon$ *with probability of at least* $1 - \delta$.

Proof Fix a target property

$$o_l = \sum_i p_i |\langle \omega_l | \psi_i \rangle|^2. \tag{6.13}$$

Conducting a single stochastic run yields the correct property in expectation, i.e.,

$$o_l = \mathbb{E} |\langle \omega_l | \tilde{\psi}_j \rangle|^2. \tag{6.14}$$

Standard concentration inequalities, like Hoeffding [58], imply

$$\mathbb{P}\left[|o_l - \hat{o}_l| \geq \epsilon\right] \leq 2e^{-2M\epsilon^2} \tag{6.15}$$

for a single property o_l. For a collection of L properties the right bound becomes

$$\delta = L \cdot (2e^{-2M\epsilon^2}) \tag{6.16}$$

by taking the union bound (also known as Boole's inequality [16]) over all L target approximations. Solving Eq. (6.16) for M results in Eq. (6.12). □

The required number of samples M scales inverse quadratically with the target accuracy ϵ— as is typical in Monte Carlo. More interestingly, M depends only logarithmically on the number of tracked properties L and is independent of the size of the system (i.e., the number of qubits of the quantum state). This logarithmic scaling can help to counteract the curse of dimensionality. For example, to approximate all $N = 2^n$ outcome probabilities of a n-qubit system up to ϵ accuracy, only roughly n/ϵ^2 samples are required.

Overall, stochastic quantum circuit simulation allows one to consider errors without the need to extend the representation of quantum states. This is achieved by viewing errors as "normal" operations during the quantum circuit simulation, which are stochastically applied during the computation, based on their probability. By generating enough samples in such a fashion, interesting properties of the quantum state can be estimated, such as, the probability of measuring specific states.

6.2 Effects on the Simulation

Stochastic quantum circuit simulation is an interesting alternative to deterministic consideration of errors. It avoids the increase in complexity from 2^n-vectors to $2^n \times 2^n$- density matrices and can be implemented in a straightforward fashion on top of an "error-free" quantum circuit simulator. Accordingly, some quantum circuit simulators already make use of this (e.g., [7, 25, 54, 89]). However, at the same time the deterministic description of the quantum state is lost, and interesting properties of the quantum state can only be approximated. In order to obtain reliable and accurate results, sufficiently many samples have to be generated.

Example 6.3 Stochastic quantum circuit simulation shall be used to accurately predict properties (e.g., the probability for measuring specific basis states) of a quantum circuit simulation. More precisely, 10,000 properties ($L = 10, 000$) shall be calculated with an error margin of $\epsilon = 0.015$ and a confidence of 97% ($\delta = 1 - 0.97$). Using Theorem 6.1, the number of required samples can be predicted:

$$M = \frac{\ln \left(\frac{2L}{\delta} \right)}{2\epsilon^2} = \frac{\ln \left(\frac{2 \cdot 10, 000}{1 - 0.97} \right)}{2 \cdot 0.015^2} = 29, 800.1 \tag{6.17}$$

Thus, at least 29,801 samples are necessary to predict 10,000 properties of the quantum state with 97% confidence and an error margin of 0.015.

The question therefore becomes how to generate the required samples—each exponential in size.

6.3 Advanced Simulation Approach

Efficiently generating the necessary number of samples is a major challenge of stochastic quantum circuit simulation. However, decision diagrams are an interesting candidate to achieve the necessary efficiency. They have already proven to be a suitable data structure to efficiently represent quantum states and also support the necessary operations for stochastic consideration of errors. Decision diagrams offer the following further potential for improvement when considering stochastic quantum circuit simulation:

6.3.1   Exploiting Concurrency

Due to their compact representation of quantum states and extensive use of sharing, decision diagrams have been unsuited for concurrent execution thus far [57]. Therefore, decision diagram-based quantum circuit simulation could not fully exploit the available hardware resources of modern (multi-core) computers, yet (in contrast to simulation approaches based, e.g., on arrays, which are easy to parallelize; see, e.g., [7, 25, 41, 54, 64, 66, 87, 89, 98, 103, 111]). Interestingly, stochastic quantum circuit simulation offers the potential of resolving the apparent conflict between optimizing memory (by using decision diagrams) and exploiting concurrency (to speed-up matrix-vector-multiplication) by other means. Now, the full potential of sharing with decision diagrams can be used, as the remaining hardware power can still be put to good use to generate more samples in parallel.

Note that this is, of course, not a new approach. Parallel execution is a well-established feature of Monte-Carlo-type approximations—and stochastic quantum circuit simulation is merely an interesting use-case. Despite this, state-of-the-art stochastic simulators such as [7, 89], however, do not seem to make use of this yet. This is most likely due to the fact that parallel executions are still more beneficial for improving, e.g., the matrix-vector-multiplications of single runs, thereby leaving no free resources for parallelizing the generation of samples. This does not constitute a problem for decision diagram-based simulation, as no parallel resources are needed for a single simulation—allowing to use the hardware power for generating samples in parallel.

6.3.2   Stacking Operations

While stochastic quantum circuit simulation can be directly implemented on top of decision diagram-based quantum circuit simulators without consideration to errors (as presented in Sect. 3.2), the compressed data structure of decision diagrams allows for further optimization of matrix-vector-multiplications by exploiting an idea first suggested in [122]. More precisely, simulating a quantum circuit boils down to applying m quantum operations $U_1, U_2, \ldots, U_m$ to an initial state $|\psi_0\rangle$, resulting in the final state $|\psi_m\rangle$, i.e.,

$$|\psi_m\rangle = U_m \cdot U_{m-1} \ldots U_2 \cdot U_1 \cdot |\psi_0\rangle. \tag{6.18}$$

This equation is usually solved from right to left as matrix-vector-multiplications are more efficient than matrix-matrix-multiplications. However, due to the compressed data structure of decision diagrams, matrix-matrix-multiplications can be notably faster than matrix-vector-multiplications for specific cases. The reason for this lies in the fact that, when working with decision diagrams, the cost of operations depends on the complexity (i.e., size) of the involved decision diagrams. Since a state vector is often more complex than a quantum operation, "stacking" (i.e., multiplying) quantum operations with each other before applying them to the quantum state can improve the simulation speed. Yet, simply stacking *all* quantum operations before applying them to the quantum state does not work, because whenever two quantum operations are multiplied, the resulting stacked operation usually becomes more complex. Therefore, the advantage of this approach relies on finding a good heuristic of how many operations are stacked, before they are applied to the quantum state. Additionally, this scheme only improves the performance when the quantum state is so complex that stacking operations is faster.

By employing this scheme during stochastic quantum circuit simulation, both aspects are considered, i.e., whenever an "intentional" quantum operation is applied, the error operations are stacked on top of it. On the one hand, this is a decent heuristic, since all operations target the same qubit, which keeps the resulting stacked operations compact. On the other hand, since the quantum circuits are simulated with consideration to errors, the quantum state naturally tends to become more complex anyway—increasing the size of the decision diagram representing it and, thus, making operations stacking more viable.

Overall, stochastic quantum circuit simulation allows one to consider errors without the exponential overhead on top of the already exponential problem of quantum circuit simulation. However, using this formalism, the deterministic description of the quantum state is lost, and in order to generate reliable results the simulation must be repeated many times. This section proposed advanced schemes to tackle the additional complexity within a decision diagram-based stochastic noise-aware quantum circuit simulator, which is extensively evaluated in the next section.

6.4 Empirical Results

The proposed stochastic noise-aware quantum circuit simulator is implemented in C++ on top of the open-source decision diagram package taken from [119, 121] and evaluated. The setup of the evaluation is presented next. Then the proposed optimizations are evaluated, the simulator is compared against state-of-the-art simulators by IBM and Atos, and, finally, the proposed stochastic simulator is compared with the deterministic simulator presented in Chap. 5.

6.4.1 Setup

The evaluation setup that was used in the previous chapter (presented in Sect. 5.4.1) is mostly reused. More precisely, for benchmarks the *Entanglement* circuit, the *Quantum Fourier Transform* (QFT) [82] with an increasing number of qubits, as well as the *QASMBench* benchmark suite [71] are used. All quantum circuits are translated into basic QASM gates (see [26]) before conducting the experiments, to ensure fair comparison.

For noise effects gate errors are approximated using depolarization with a probability of 0.1%, amplitude damping error (T1) with 0.2% probability, and phase-flip error (T2) with

0.1% probability. These probabilities are doubled for multi-qubit operations and applied to a qubit whenever it is used in an operation. For each experiment 30,000 samples are generated to predict the final state (according to Theorem 6.1, this translates to tracking 10,941 properties with an error margin of 0.015 and a confidence of 97%).

The experiments were carried out on a system with 96 cores running at a clock frequency of 2.2 GHz and 1.5 TB of R,AM. As (industry-grade) state-of-the-art representatives of stochastic quantum circuit simulators the *LinAlg* simulator of Atos' *QLM* [7] and the *statevector* simulators of IBM's Qiskit are used.

The QLM simulator ran directly on the system, while Qiskit and the proposed simulator are ported to the machine using Docker [75], which was used since its virtualization overhead is negligible [36], ensuring a fair comparison. Finally, a timeout of half an hour (1800 seconds) was enforced for all experiments and, in the following summary, only the most relevant benchmarks are listed, i.e., benchmarks that could be simulated by *all* approaches in less than a few seconds, and benchmarks that no simulator could simulate within the given time limit are omitted.

6.4.2 Comparison Between Implementations

In a first series of evaluations the proposed optimization is evaluated. To this end, the concepts are implemented in a basic fashion (i.e., as presented in Sect. 6.1.2) and one where the optimization presented in Sect. 6.3.2 is implemented.[2]

The results of the comparison between the two implementations are provided in Table 6.1. In the table the name of the benchmark, the number of qubits (#Q), the number of gates (#G), the runtime for the basic, the optimized implementation, as well as the runtime improvements with the optimization compared to the basic implementation are listed.

The results show that the optimizations improve the runtime considerably for all circuits. While the performance improvements vary from circuit to circuit, runtime improvements from 28.62% up to 56.06% can be reported.

6.4.3 Comparison to Related Work

In a second series of evaluations, the optimized version of the proposed simulator is compared against the *LinAlg* simulator of Atos' *Quantum Learning Machine* (QLM) [7], as well as the *statevector* simulators of IBM's *Qiskit* [89]. In addition to running all experiments with decoherence and depolarization errors—here the scenario with depolarization errors only is also considered. In order to evaluate how different noise models affect the performance.

---

[2] Note that concurrent execution, as proposed in Sect. 6.3.1, was used in both implementations, since the performance improvement of parallelization is straightforward.

Table 6.1 Optimization for stochastic simulation

Benchmark	#Q	#G	Basic [s]	Optimized [s]	Runtime Imp. (%)
basis_trotter	4	1626	28.79	**13.58**	52.83
qaoa	6	270	11.46	**8.18**	28.62
vqe_uccsd	6	2282	83.23	**54.65**	34.34
vqe_uccsd	8	10808	1214.14	**849.64**	30.02
ising	10	480	201.48	**139.51**	30.76
sat	11	679	1439.97	**981.26**	31.86
seca	11	282	1619.35	**1119.47**	30.87
cc	12	49	255.70	**181.64**	28.96
multipler	15	574	736.65	**317.06**	56.96
bigadder	18	284	151.76	**79.11**	47.87
bv	14	43	85.15	**41.37**	51.42
bv	19	58	633.55	**361.19**	42.99

Table 6.2 summarizes the results of the experiments.[3] For each experiment the name of the benchmark, the number of qubits (#Q), number of gates (#G), as well as the required runtime for simulating it with the QLM, Qiskit, and the decision diagram-based solution in seconds is listed. The results of Atos' QLM simulator are omitted for the QASMBench benchmarks, due to compatibility issues.

The results clearly show where noise-aware quantum circuit simulation with decision diagrams greatly improves current state-of-the-art. The stochastic decision diagram-based simulator outperforms the considered industry-grade solutions for many quantum applications, often even in the order of magnitudes.

6.4.4   Comparison Between Deterministic and Stochastic Scheme

Finally, a third series of evaluations considers the difference in the performance of both decision diagram-based simulation schemes, i.e., the proposed stochastic simulator and the deterministic simulator proposed in the previous chapter. To this end, both approaches are directly compared. To gain more insight into the behavior of both simulation schemes, the resource consumption of both simulators is also tracked, in addition to the runtime. Table 6.3 provides the obtained results. Again, the table contains the name of the benchmark, the

---

[3] When simulating larger instances of the QFT circuit with the QLM the simulation terminated with an internal error. This is denoted in the table as "error".

Table 6.2 Comparison to state-of-the-art stochastic quantum circuit simulators

Noise model	Entanglement circuit				QFT circuit				QASMBench benchmarks			
	# Q/#G	QLM	Qiskit	Prop.	# Q/#G	QLM	Qiskit	Prop.	Benchmark	# Q/#G	Qiskit	Prop.
Decoherence and Depolarization	21/21	35.46	1130.81	**6.10**	9/192	370.69	14.37	**3.97**	basis_trotter	4/1626	78.27	**13.58**
	22/22	71.83	>1800.00	**6.47**	10/241	650.46	18.12	**4.30**	vqe_uccsd	6/2282	97.51	**54.65**
	23/23	165.92	–	**6.28**	11/291	error	20.79	**4.34**	vqe_uccsd	8/10808	477.82	849.64
	24/24	232.98	–	**6.05**	…	–	…	…	ising	10/480	**18.29**	139.51
	25/25	456.56	–	**6.14**	19/886	–	1436.19	**7.88**	sat	11/679	39.39	981.26
	26/26	920.62	–	**6.48**	20/977	–>	1800.00	**9.27**	seca	11/282	12.28	1119.47
	27/27	>1800.00	–	**6.57**	…	–	–	…	multiply	13/124	10.11	4.81
	…	–	–	…	61/9239	–	–	153.58	multiplier	15/574	**90.79**	317.06
					62/9551	–	–	174.75	bigadder	18/284	478.91	**79.11**
					63/9856	–	–	194.19	cc	18/73	172.97	>1800.00
					64/10177	–	–	192.70	bv	19/58	266.16	361.19
Depolarization	22/22	21.43	1675.15	**5.93**	15/548	703.25	43.86	**5.28**	basis_trotter	4/1626	24.85	**10.89**
	23/23	43.78	>1800.00	**5.96**	16/624	967.45	69.04	**5.50**	vqe_uccsd	6/2282	40.95	**38.01**
	24/24	108.05	–	**6.15**	17/704	error	117.71	**5.97**	vqe_uccsd	8/10808	200.37	393.76
	25/25	226.07	–	**6.31**	18/793	–	245.27	**6.15**	ising	10/480	**4.77**	81.01
	26/26	446.78	–	**6.37**	19/886	–	517.64	**6.66**	sat	11/679	11.18	**10.00**
	27/27	1452.45	–	**6.56**	20/977	–	1091.63	**7.19**	seca	11/282	4.27	**4.18**
	28/28	>1800.00	–	**6.61**	21/1080	–	>1800.00	**7.44**	multiply	13/124	**3.85**	4.94
	…	–	–	…	…	…	–	…	multipler	15/574	47.10	**5.22**
	62/62	–	–	**11.34**	62/9551	–	–	**89.38**	bigadder	18/284	120.07	**5.26**
	63/63	–	–	**11.42**	63/9856	–	–	**92.84**	cc	18/73	107.31	9.91
	64/64	–	–	**11.47**	64/10177	–	–	**100.33**	bv	19/58	183.50	**5.55**

Table 6.3 Comparison between deterministic and stochastic scheme

Benchmark	#Q	#G	Det. T. [s]	Stoch. T. [s]	Det. CPU-T. [s]	Stoch. CPU-T. [s]	Det. Mem. [MB]	Stoch. Mem. [MB]
basis_trotter	4	1626	**0.41**	13.58	**0.27**	1149.50	100.54	1293.60
qaoa	6	270	**0.75**	8.18	**0.62**	675.69	100.37	1230.77
vqe_uccsd	6	2282	**4.01**	54.65	**3.88**	4811.25	100.62	1692.11
vqe_uccsd	8	10808	1301.89	**849.64**	**1298.23**	75407.87	118.75	1808.42
qpe	9	150	6.24	**4.45**	**6.11**	327.10	128.36	1278.00
qftRandom	10	241	73.23	**4.30**	**72.90**	303.16	194.25	1292.18
qftRandom	11	291	1382.62	**4.34**	1378.64	320.54	506.52	1332.33
qftRandom	12	346	>1800.00	**4.77**	–	342.15	506.48	1347.92
qftRandom	13	410	–	**4.88**	–	353.56	–	1454.21
multiply	13	124	303.94	**4.81**	**302.93**	339.08	170.20	1388.46
entanglement	31	31	953.95	**7.11**	950.96	537.16	1010.15	2220.31
entanglement	32	32	1375.30	**7.12**	1371.23	552.63	1010.29	2109.41
entanglement	33	33	>1800.00	**8.21**	–	485.61	1906.23	2311.31
entanglement	34	34	–	**9.17**	–	648.61	–	2861.76

number of simulated qubits (#Q), the number of gates (#G) , as well as the runtime in seconds for both simulation approaches. Furthermore, the CPU time and the maximum memory usage in MB are provided.

Looking at the absolute runtime, the deterministic simulator is only faster for smaller quantum circuits, whereas the stochastic approach scales much better and manages to simulate circuits that are not feasible with the deterministic approach. Comparing the resource consumption shows a more diverse picture, with the stochastic simulator often using more CPU time and always using much more memory than the deterministic simulator. Nevertheless, the increased performance of the stochastic simulator comes with sacrificing the deterministic description of the quantum state, and hence, a potential loss of accuracy. In practice, however, this drawback is often negligible, as noise-aware quantum circuit simulation is only an approximation of real quantum computers in the first place, and, as shown in Theorem 6.1, the error margin of the approximated results can be reduced to the desired accuracy by generating more samples. Hence, while having a deterministic approach as a baseline for an exact consideration of noise/errors, the (much more efficient and scalable) stochastic approach is more suited for the simulation of practically relevant instances.

6.5   Conclusion

This chapter demonstrated the potential of decision diagrams for noise-aware stochastic quantum circuit simulation. To this end, the necessary formalism was introduced and a rigorous mathematical model to calculate the required number of samples was presented. Then, an optimized decision diagram-based stochastic quantum circuit simulator was proposed that exploits concurrency and applies operations in a more efficient manner and implemented in C++.

In an extensive evaluation, the decision diagram-based simulator was then empirically evaluated. More precisely, the usefulness of the proposed optimizations was investigated, the implemented simulator was compared against state-of-the-art solutions from IBM and Atos, and, finally, the solution was compared against the deterministic simulator proposed in the previous chapter.

The findings confirmed the advantages of the proposed optimization, showed considerable improvements compared to the industry-grade simulators for many circuits, and even showed considerable speed-ups compared to the deterministic solution proposed in the previous chapter. The implementation used for the evaluation is available as open-source under the MIT license under https://github.com/cda-tum/mqt-ddsim.

Part IV
Application

Error Correction Framework

The previous part demonstrates the potential of decision diagrams for noise-aware quantum circuit simulation. These simulators that have been proposed, implemented, evaluated, and published as open-source allow users to examine how their quantum applications would behave when executed on a real quantum computer. A class of quantum algorithms for which such a noise-aware quantum circuit simulation is essential is quantum error correction schemes. These schemes correct errors that occur in real quantum hardware, and as such are believed to be essential in creating an error-prone environment for executing large quantum applications [88].

But error correction is not easy, since qubits are much more complex than classical bits, a much wider variety of errors is possible. Worse still, directly measuring qubits destroys any quantum information (see Sect. 2.1.1); thus, it is not easy to even identify errors. Nevertheless, considerable developments have been made in quantum error correction, and today there are several quantum correction schemes available, such as CSS codes [21, 97, 101], stabilizer codes [45], surface codes [17, 28, 67], Bacon-Shor codes [8], adaptive codes [38, 92] and many more.

However, despite this progress, most of the corresponding work in this domain still relies heavily on manual labor and/or is based on theoretical results only. Methods (or implementations thereof) that take a quantum circuit and *automatically* extend it with a correspondingly chosen error-correcting scheme (similar to a compiler that translates a given quantum algorithm to a corresponding hardware-applicable realization) are hardly available. This may be due to the fact that so far the focus has clearly (and understandably) been on the development of concepts for error-correcting codes and proof-of-concept implementations on selected hardware models. Additionally, realizing such functionality is not straightforward. For example, the decoding and correction steps are exponentially hard [59] and, although

© The Author(s), under exclusive license to Springer Nature Switzerland AG 2025
T. Grurl et al., *Noise-Aware Quantum Circuit Simulation with Decision Diagrams*,
Synthesis Lectures on Engineering, Science, and Technology,
https://doi.org/10.1007/978-3-031-71036-0_7

this is addressed in recent work (such as [11]), they remain complex. Furthermore, applying quantum operations to qubits protected by an error-correcting code is often not directly possible, but requires specific routines (as shown by the no-go theorem [32]).

As a consequence, evaluations on whether an error-correcting code is useful in different scenarios are cumbersome and, hence, often done with rather selected use-cases only. Because of this, there are no comprehensive case studies in which the usefulness of error-correcting codes is evaluated with respect to different circuits, how they are configured, and on what hardware model they are applied.

This chapter (based on [53]) addresses this issue by presenting an open-source framework, which supports engineers and researchers in the task of evaluating error-correcting codes. The framework allows to *automatically* apply error correction schemes to a given quantum circuit, followed by an automatic noise-aware quantum circuit simulation. To this end, different error-correcting codes are implemented and existing methods are utilized for simulating quantum circuits. In this way, a framework is created that allows quantum error-correcting codes to be easily analyzed, with minimal manual effort. The proposed framework is implemented in such a modular way that it can be readily extended for new quantum error-correcting codes or different simulation styles. Case studies show that the proposed framework allows for efficient evaluations of quantum error-correcting codes depending on varying properties. Last but not least, the case studies also demonstrate the necessity for noise-aware quantum circuit simulation.

The remainder of this chapter is organized as follows: In Sect. 7.1, the proposed framework is motivated by revisiting the basics of quantum error correction and illustrating current problems. Section 7.2 presents the proposed framework, followed by Sect. 7.3 demonstrating the application and usefulness of the framework. Finally, Sect. 7.4 concludes the chapter.

7.1   General Concept and Motivation

As reviewed in Sect. 2.2, quantum computers are plagued by noise, drastically limiting their usefulness in the real world. Quantum error correction addresses this problem, which makes it an essential part of building scalable and resilient quantum hardware. This section reviews the main ideas of the corresponding concepts and provides the motivation for an easy-to-use framework for quantum error-correcting codes.

7.1.1   Quantum Error Correction

In order to illustrate quantum error correction, first classical error correction is revisited, as it serves the same purpose. In the classical world, error correction is achieved by encoding information beyond its theoretical minimum. The redundant information is then used to identify and (possibly) correct errors. The following example illustrates this:

Example 7.1 Suppose a sender needs to transfer a single (classical) bit 0 to a receiver at a location elsewhere. Unfortunately, the transmission channel is susceptible to bit-flip errors, which is why both sender and receiver agree to protect the bit, using the three-bit repetition code. For this code, each bit of information is encoded by tripling it, i.e., 0 → 000 and 1 → 111. Hence, the one-bit message 0 is encoded into the codeword 000 before being transmitted to the receiver. During transmission, a bit-flip error occurs (distorting the message, e.g., to 001), the receiver could, through majority voting, still infer that the sent message most likely was 000 and could therefore correctly restore the original message.

In the quantum world error correction is done similarly, but some properties of quantum mechanics have to be considered, which makes the process more complex. A key difference between the quantum and the classical world is that measurements in the quantum world affect the observed system. So, while the classical system can be measured without risk of compromising the encoded information, special care must be taken in quantum error correction so as to not destroy information by measuring it. This is worsened by the no-cloning theorem [116], which asserts that it is not possible to clone (i.e., copy) arbitrary quantum states. It is therefore not possible to simply clone qubits before measuring them to keep their information intact.

Instead, to add redundancy in the quantum world, the Hilbert space, in which the information is encoded, is expanded—effectively distributing the information of a single qubit among more qubits [97]. This is illustrated using the three-qubit bit-flip code, which allows detection and correction of single-qubit bit-flip errors. A single qubit is encoded by entangling it with two ancillary qubits (this can be achieved by two CNOT operations), i.e.,

$$|\psi\rangle \rightarrow |\psi_L\rangle \tag{7.1}$$

$$\alpha_0 \cdot |0\rangle + \alpha_1 \cdot |1\rangle \rightarrow \alpha_0 \cdot |000\rangle + \alpha_1 \cdot |111\rangle. \tag{7.2}$$

After encoding the information of the state $|\psi\rangle$, it is distributed among the three-qubit state $|\psi_L\rangle$. More precisely, the information of $|\psi\rangle$ (encoded into the two-dimensional Hilbert space

$$\text{span}\{|0\rangle, |1\rangle\} \tag{7.3}$$

is now encoded into the eight-dimensional Hilbert space

$$\text{span}\{|000\rangle, |001\rangle, |010\rangle, |011\rangle, |100\rangle, |101\rangle, |110\rangle, |111\rangle\}. \tag{7.4}$$

This eight-dimensional Hilbert space can be split into four subspaces

$$\mathcal{C} = \text{span}\{|000\rangle, |111\rangle\}, \mathcal{F}_1 = \text{span}\{|001\rangle, |110\rangle\}, \tag{7.5}$$

$$\mathcal{F}_2 = \text{span}\{|010\rangle, |101\rangle\}, \text{ and } \mathcal{F}_3 = \text{span}\{|100\rangle, |011\rangle\}, \tag{7.6}$$

where the subspace $\mathcal{C}$ represents the *logical code space* (indicating that the system is in a valid state) and the subspaces $\mathcal{F}_1, \mathcal{F}_2$, as well as $\mathcal{F}_3$ represent *logical error spaces* (each indicating a bit-flip error). That is, $\mathcal{F}_1$ indicates that a bit-flip error occurred in the first qubit, $\mathcal{F}_2$ indicates a bit-flip in the second qubit, and $\mathcal{F}_3$ indicates a bit-flip in the third qubit. By using a special kind of measurement, it is possible to infer which qubits are equal (an example realization of such an indirect measurement is provided in Sect. 7.2.1). With this knowledge, it is possible to infer in which of the subspaces $\mathcal{C}, \mathcal{F}_1, \mathcal{F}_2$ or $\mathcal{F}_3 \, |\psi_L\rangle$ resides. Note that this measurement does not reveal any information on whether the encoded qubit is 0 or 1 and, therefore, does not change the encoded qubit.

Example 7.2 Imagine a scenario where a sender wants to transfer the single qubit

$$|\phi\rangle = \sqrt{1/3} \cdot |0\rangle + \sqrt{2/3} \cdot |1\rangle \tag{7.7}$$

to a receiver located elsewhere. The transmission channel is prone to bit-flip errors, so the sender and receiver agree to protect the qubit using the three-qubit bit-flip code (presented above). The sender encodes the qubit

$$|\phi\rangle \to |\phi_L\rangle \tag{7.8}$$

$$\sqrt{1/3} \cdot |0\rangle + \sqrt{2/3} \cdot |1\rangle \to \sqrt{1/3} \cdot |000\rangle + \sqrt{2/3} \cdot |111\rangle \tag{7.9}$$

and sends $|\phi_L\rangle$ to the receiver.

During the transmission, the first qubit flips, resulting in the state

$$\sqrt{1/3} \cdot |100\rangle + \sqrt{2/3} \cdot |011\rangle. \tag{7.10}$$

Before using the qubit, the receiver measures whether all physical qubits that make up $|\phi_L\rangle$ are equal. The measurement shows that the first qubit is different from the other two qubits, and hence, the receiver infers that $|\phi_L\rangle$ resides in the subspace $\mathcal{F}_3$, indicating that a bit-flip error has occurred in the first qubit. The receiver therefore applies an X operation to the first qubit of $|\phi_L\rangle$—restoring it back to

$$|\phi_L\rangle = \sqrt{1/3} \cdot |000\rangle + \sqrt{2/3} \cdot |111\rangle \tag{7.11}$$

Afterwards, the receiver can decode $|\phi_L\rangle$ to $|\phi\rangle$ and use it in further calculations.

While the concepts presented above allow handling bit-flip errors (something that may be sufficient in the classical world), the quantum world allows for a substantially larger continuum of errors. For example, recall that the depolarization error, introduced in Sect. 2.2.2, leaves a qubit in a completely random state—this cannot be corrected using bit-flip correction only. Even worse, qubits also contain phase information, which has no classical counterpart at all and which must also be protected. Fortunately, it turns out that it is sufficient to correct only a subset of errors to correct *all* possible unitary errors that can occur in quantum computers [68]. In practice, it suffices to consider bit-flip and phase-flip errors to have a universal error-correcting code [93]. In 1995 Peter Shor developed the first quantum error-correcting code that accomplished just that [97]. Since then, many quantum correction schemes have been developed, such as CSS codes [21, 97, 101], stabilizer codes [45], surface codes [17, 28, 67], Bacon-Shor codes [8], adaptive codes [38, 92] and many more.

7.1.2   Motivation

As shown above, quantum error correction tackles the problem of errors in quantum computations, and various error correction schemes are available. However, using these schemes is not as straightforward as Example 7.2 suggests. Noisiness is not restricted to specific erroneous quantum channels, but affects *all* operations, including those operations necessary for error correction.

Example 7.3 The effects of noise are different depending on the simulated hardware model. To illustrate this, the fidelity, as defined in Definition 2.6, is used to express "correctness" of state $|0\rangle$ depending on the probability of bit-flip error occurring in three different scenarios—without any kind of error protection, with the ideal bit-flip error protection presented in Example 7.2, and with a more realistic bit-flip error protection. In the latter case, the operations necessary to realize the bit-flip error correction are also affected by noise (this is reasonable to assume, as discussed in Sect. 2.2).

Figure 7.1 shows that without any kind of error correction (red curve), the fidelity of the encoded state more or less linearly declines with increasing error probability. Using the ideal bit-flip error protection (blue curve) increases the fidelity, as long as the error probability stays below 50%. Finally, in the realistic case (brown curve), the fidelity severely drops compared to the case where no error correction is applied at all.

The example illustrates that naively applying error correction schemes to quantum circuits can lower the fidelity (as proven by the threshold theorem [4, 67, 69]). The optimal error correction scheme depends on the properties of the quantum circuit as well as the noisiness of the considered quantum computer. Additionally, it is necessary to optimally tailor the error correction scheme to the specific use-case. Thus, a thorough evaluation is needed.

However, while there exists a lot of research on quantum error correction (e.g., [17, 21, 28, 38, 45, 92, 97, 101]), the focus clearly (and understandably) is on the development of corresponding concepts for error-correcting codes and proof-of-concept implementations on selected quantum circuits. As illustrated above, however, evaluating error-correcting codes requires a broader consideration—thus far, involving extensive research and tedious manual implementation of said schemes. To address this problem, a framework is proposed that automates the process of applying error correction to circuits and also allows for evaluation using noise-aware quantum circuit simulation.

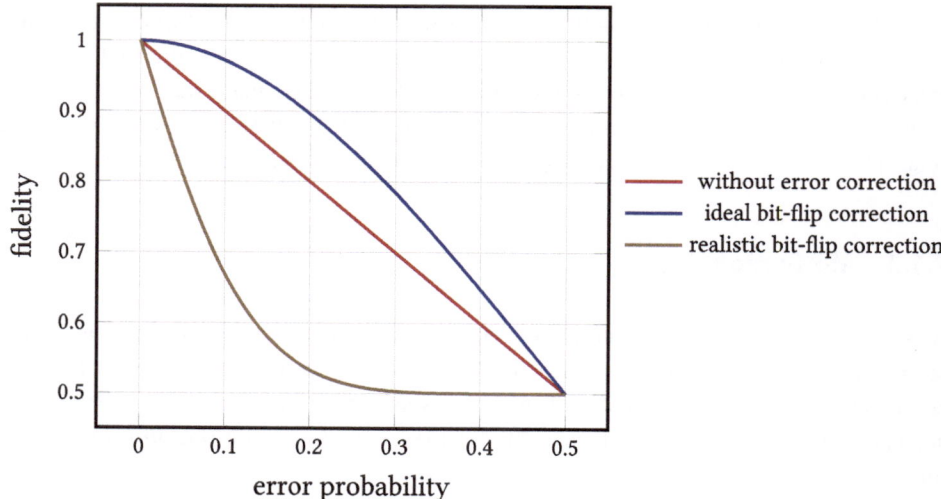

Fig. 7.1 State fidelity with increasing error probability

7.2   Proposed Framework

The proposed framework is supposed to support the process of evaluating error correction schemes. To this end, it not only allows one to apply error-correcting codes to a given quantum circuit, but also supports the entire circuit preparation and simulation process. More precisely, the framework (1) automates the circuit compilation flow (including the application of error-correcting codes), (2) provides quantum circuit simulation with noise-aware quantum circuit simulators, and (3) is flexible and easily extensible for new use-cases.

Note that while circuit compilation and simulation are necessary parts of such a framework, there already exist available solutions (e.g., [25, 42, 50, 54, 89]). Therefore, the proposed framework is realized in such a manner that it can be coupled with these solutions. In the remainder of this section, the proposed framework is presented.

7.2.1   Compiler

During the compilation process, the quantum circuit is prepared for circuit simulation or execution on a real quantum computer. Here, the main functionality provided by the framework is focused, namely the automatic application of error-correcting codes to the quantum circuit. For this, the fact is exploited that, while there exist many different error correction schemes, all of them must in practice realize the same steps when applied to a quantum circuit. These steps are as follows:

1. Qubit Encoding

The basic idea of all error-correcting codes is that information is encoded in such a way that it allows the detection and correction of errors that occur. Although encoding depends on the respective approach, all schemes have in common that information is distributed among multiple qubits. At the end of this step, all qubits within the logical circuit are encoded into logical qubits consisting of multiple physical qubits.

2. Operation Encoding

Having a quantum circuit with encoded qubits, the next step is to adjust the operations. More precisely, the originally intended operations are mapped to logical operations, i.e., operations that are functionally equivalent but respect the applied qubit encoding. The mapping of each operation is not always trivial [32] and depends on the error correction scheme used. As a consequence, only specific quantum operations are supported for each error-correcting code.

3. Error Correction

Error detection and correction routines are added to the circuit. This includes syndrome extraction followed by some corrective operations on the logical qubits—depending on the extracted syndrome. The optimal frequency of executing the detection and correction routine, i.e., after how many quantum operations this routine is applied to a logical qubit, varies and depends on several factors, e.g., the hardware specifications of the designated quantum computer or the error-proneness of the application. To accommodate this, the frequency of the detection and correction step can be freely adjusted.

4. Qubit Decoding

In the last step, the logical qubit is decoded back to a single physical qubit for measurement. For convenience, this is realized in such a way that the ordering of the output qubits is equal to the order of the qubits in the original circuit.

The following example illustrates the process of applying an error-correcting code using the bit-flip code presented in Example 7.2.

Example 7.4 In Fig. 7.2, a circuit is presented without error correction (Fig. 7.2a) and with bit-flip error correction (Fig. 7.2b), which is a simplified version of Shor's error correction proposed in [97]. The color-coding in the circuit with error correction represents during which step the respective parts are generated. During encoding phase, each qubit of the original circuit is encoded into a logical qubit by entangling it with two ancillary qubits. Next, every quantum operation is copied for each qubit that makes up the logical qubit,

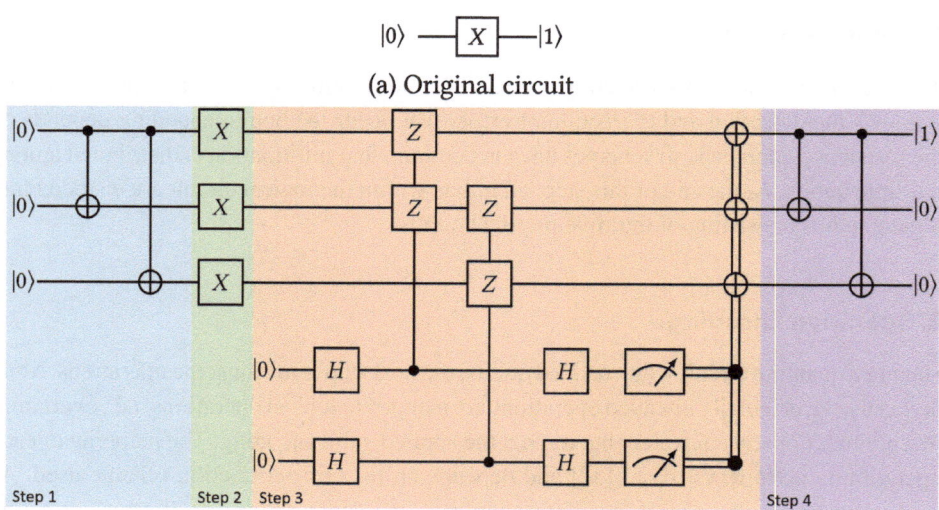

(b) Circuit with bit-flip error correction

Fig. 7.2 State Applying the bit-flip error-correcting code to a circuit

e.g., an X operation in the original circuit is mapped to an X operation onto each qubit that makes up the respective logical qubit. Then, error detection and correction is added. For error detection, a projective measurement is added, which is realized by two ancillary qubits and four controlled Z operations. The ancillary qubits then contain the syndrome, which indicates whether a bit-flip error has occurred. If that is the case, the flipped qubit is corrected by applying an X operation to it. Finally, the qubit is decoded so that it can be measured.

Since these steps are necessary for all error-correcting codes, the framework provides templates for their implementation. Due to this modular design, the framework can be easily extended, either by modifying the already available error-correcting codes or by adding new ones. Currently the framework supports four error-correcting codes ranging in size to different error correction types, namely the Shor Code [97] (as this is the first code that corrects arbitrary single-qubit errors), the Laflamme Code [70] (as this is the smallest possible code protecting against arbitrary single-qubit errors), the Steane code [101] (as it is a well-studied code used in other experiments such as [92]), and a surface code [39, 117] (due to their popularity in the current research on quantum error-correcting codes).

7.2.2 Simulator

An essential part of the framework is the ability to directly simulate circuits with (and without) error correction and different hardware models. For this task, the decision diagram-based quantum circuit simulator developed in the previous part can be used. In addition to that, several simulation approaches have recently been introduced in the literature (e.g., [25, 42, 50, 54, 89]). Some of these are available as open-source and hence can be readily integrated into the proposed framework. By this, access to multiple simulation styles is gained, which come with their advantages/drawbacks, and the most appropriate simulator can be selected for each use-case.

Example 7.5 To illustrate these differences between noise-aware quantum circuit simulation approaches, the entanglement circuit (which constructs the GHZ state over all qubits) is simulated with an increasing number of qubits. During all simulations, depolarization noise is applied with 0.001% probability for single qubit gates and 0.002% for multi-qubit gates, which is applied to a qubit whenever it has been used. The experiment was carried out with an array-based density matrix simulator [89], an array-based stochastic simulator [89], a stabilizer-based simulator [89], as well as the decision diagram-based (DD-based) density matrix (deterministic) simulator (proposed in Chap. 5) and the decision diagram-based (DD-based) stochastic simulator (proposed in Chap. 6). For each simulation 30,000 shots have been performed. In Fig. 7.3 the simulation times are plotted as a function of the number of simulated qubits.

The plot shows that the runtime of both array-based simulators and the decision diagram-based density matrix simulator grows exponentially, with increasing number of qubits, while the stochastic decision diagram-based simulator and the stabilizer-based simulator exhibit

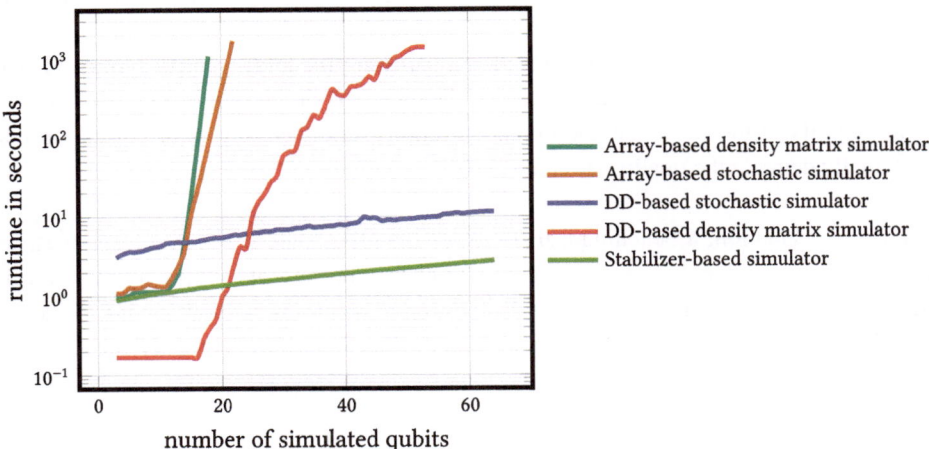

Fig. 7.3 Runtime of noise-aware quantum circuit simulation depending on the simulation style

polynomial runtime. However, as shown in the previous chapters, decision diagram-based simulators are *not always* faster than array-based simulators. Furthermore, while density matrix simulators perform worst in runtime, they offer a *full* and deterministic description of the final state. In contrast, the stochastic solutions only approximate the final quantum state. The stabilizer-based simulator performs best for this circuit, but only supports specific quantum operators (i.e., the Clifford gate set), which limits its useability.

This small experiment shows the advantage of having a variety of different quantum circuit simulators available for experiments. In the framework this is implemented in such a way that the simulator can be easily switched to the most fitting one. To this end, quantum circuit simulators can be directly integrated into the framework, as long as they support a Python interface. Alternatively, the transformed circuit can also be exported as an Open-QASM circuit, which is the de facto standard format for quantum circuits [26].

7.3   Application and Demonstration

The usefulness of the error correction framework is demonstrated by considering the Steane code [101] as a well-known representative of an error-correcting code. For this code it is evaluated in which scenarios the error correction actually improves the reliability and in which it does not.[1] To this end, the effects of three different properties are evaluated:

1. Size of the considered circuit: error correction is a costly procedure, as illustrated in Fig. 7.2. Whether error correction improves the reliability of a circuit substantially depends on its size.
2. Frequency of error correction: Detecting and correcting errors is a costly operation. The appropriate frequency with which this routine is performed is critical when applying an error-correcting code.
3. Assumed hardware model: The considered hardware model (i.e., error types and error probability) also affects the usefulness of the quantum circuit. For example, if the error probability is too high, the additional operations introduced by error correction may actually degrade the quality of the results.

For the evaluation, a benchmark that creates the GHZ state between all qubits (taken from [90]) is considered. This benchmark was chosen because its simplicity makes it the ideal candidate for testing the Steane code in different scenarios, without having to account for characteristics within the data. Besides that, the benchmark can be easily varied in size,

---

[1] Using the proposed framework, similar considerations can be done with other codes as well.

and the created states are strongly correlated, thereby making (uncorrected) errors easily noticeable. If not stated otherwise, the following default parameters have been used in the evaluation:

- Depolarization noise (applied whenever a qubit is used) has been considered with an error probability of 0.001%.
- Error correction and detection has been applied to a logical qubit prior to the measurement and whenever it has been used 500 times.
- The entanglement benchmark is used with five qubits and 10,000 dummy operations added to the end of the circuit (the dummy operations do not change the result but are affected by noise).

For each property discussed above, this default use-case is adapted to evaluate the respective property. That is, for (1) the entanglement circuits were simulated with an increasing number of dummy operations, for (2) the entanglement circuits were simulated with a decreasing rate of error correction steps, and, finally, for (3) the benchmark was simulated with increasing error probability not only with depolarization noise (mimicking gate errors) but also with amplitude damping errors (mimicking decoherence errors). In the latter case the entanglement benchmark was used with two qubits (instead of five), since a stochastic array-based simulator was used for simulating the amplitude damping noise.

The proposed framework was used to encode all circuits with the Steane code. Afterwards, the circuits were simulated using Qiskit [89]. More precisely, the stabilizer-based simulator was used when considering depolarization errors, and the state vector simulator when considering amplitude damping errors (as these cannot be simulated with the stabilizer simulator). All experiments were simulated stochastically with 2000 shots. The simulations yielded the probability distribution for measuring specific basis states, which was used to calculate the "classical" fidelity (also known as the Hellinger coefficient [13]) between the expected probability distribution (i.e., the received probability distribution when no errors occur during the computation) and the actual distribution received by the noise-aware quantum circuit simulation. This distance was chosen as a metric over others, such as the fidelity between the quantum states [65] or the average fidelity, as used for example in [43]), since the aim of the evaluation was to evaluate how a noisy quantum computer affects the "correctness" of the measured final states for *specific* quantum algorithms.

For these evaluations, 285 simulations were conducted and 137 times error correction was applied to the quantum circuits.

In Figs. 7.4, 7.5 and 7.6 the results of the experiments are summarized. Figure 7.4 depicts the fidelity with increasing depth of the entanglement circuits. It shows that the usefulness of error correction increases with the gate depth. Conducting error correction only pays off when the circuit is of sufficient depth. Next, in Fig. 7.5 the fidelity is plotted with decreasing error correction frequency. Contrary to intuition, it is not advantageous to correct errors as

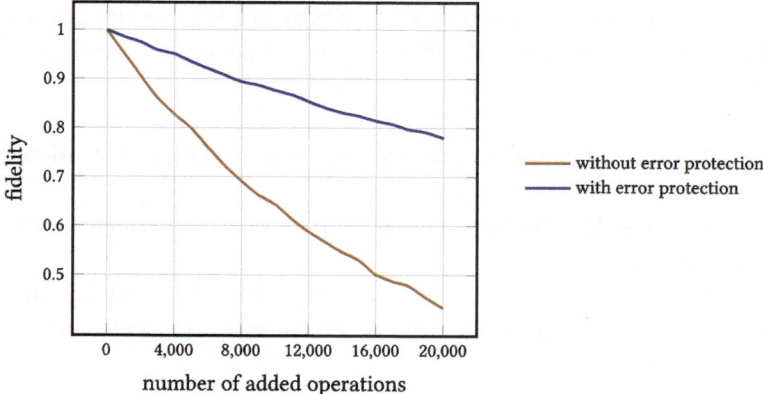

Fig. 7.4 Evaluation of the circuit size (use-case 1)

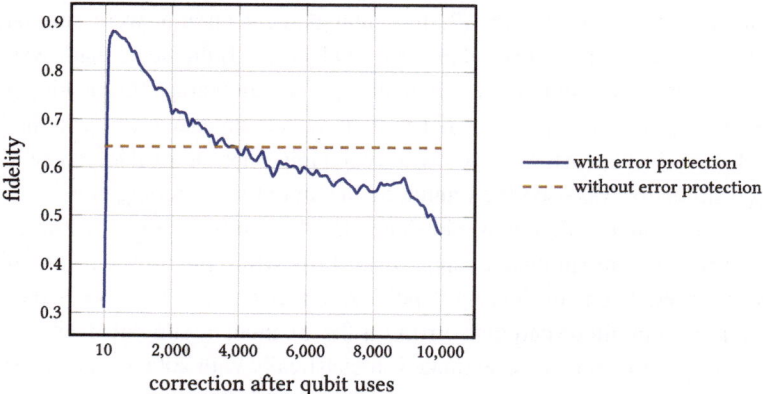

Fig. 7.5 Evaluation of the error correction frequency (use-case 2)

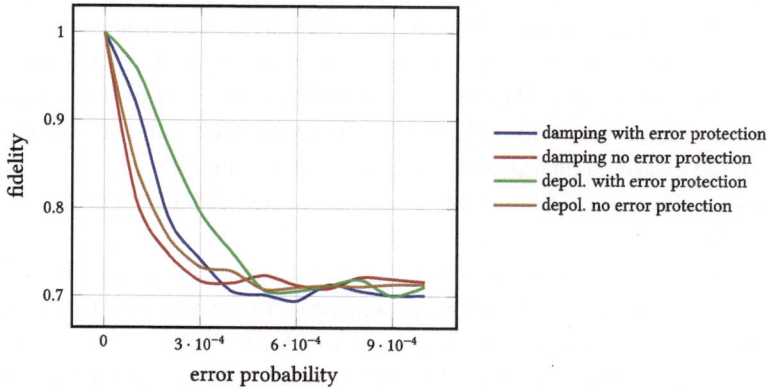

Fig. 7.6 Evaluation of the hardware model (use-case 3)

often as possible during computation. This can be explained by the fact that, as discussed above, error correction is expensive and adds a substantial amount of extra operations. Hence, it has a positive effect if error correction application and resulting overhead are traded-off (a task for which the proposed framework offers a very helpful tool). Finally, Fig. 7.6 depicts the fidelity with increasing (physical) error probability and different types of errors. The plot shows that error correction is useful only as long as the error probability stays low. Additionally, the simulated noise type noticeably affects the effectiveness of the error correction scheme. Thus, depolarization errors could be handled more effectively than amplitude damping errors.

Obviously, these demonstrations only provide a snippet of what kind of evaluations can be performed with the proposed framework. But it clearly demonstrates the usefulness of the approach. While until now, applying error correction to circuits and assessing their effects involved tedious manual work. Now, the proposed framework automates this task, making quantum error correction more widely applicable and simplifying analysis.

7.4 Conclusion

Quantum error-correcting codes are an essential part of building resilient and scalable quantum hardware. However, currently most of the corresponding work in this domain is heavily dependent on manual labor and/or is based on theoretical results only. This problem was addressed in this chapter.

To this end, a framework was proposed that automates the process of applying quantum error correction and also allows noise-aware simulation of the protected circuits. The advantages and usefulness of the proposed framework were demonstrated by evaluating the reliability of the Steane error-correcting code with respect to different parameters.

The proposed framework was implemented in a modular fashion, so that it can be easily configured and extended for new error-correcting codes and is easy to use, making quantum error correction more broadly applicable. The respective source code of the framework is available under MIT license at https://github.com/cda-tum/mqt-qecc/.

Last but not least, the framework demonstrates the need for noise-aware quantum circuit simulators, showing the usefulness of the tools developed in the pursuit of this book.

Part V
Conclusion

Conclusion

8

Noise-aware simulation of quantum circuits on classical hardware is an important task for understanding, developing, and evaluating quantum algorithms. However, considering noise effects during quantum circuit simulation makes an already exponentially hard problem even more complex—severely limiting corresponding approaches. Decision diagrams are a promising tool to tackle this complexity, as they offer a compact data structure for representing quantum functionality that is often below the exponential size of a straightforward representation. Although originally developed to solve (exponential) problems in the world of classical design automation, they have since proven their usefulness for solving design tasks in the quantum world as well. Thus far, however, their potential for noise-aware quantum circuit simulation remained largely unexplored. This was changed with this book, which comprehensively explores the potential of decision diagrams for noise-aware quantum circuit simulation.

The logical starting point of this investigation was decision diagram-based quantum circuit simulation *without* any consideration of noise. Thus, the state-of-the-art of quantum decision diagrams was reviewed and evaluated first. To this end, the cost of multiplication and addition of decision diagrams—essential operations for quantum circuit simulation—was evaluated first and it was shown that adding decision diagrams can be more costly than multiplying them. Afterwards, the performance of a state-of-the-art decision diagram-based quantum circuit simulator was compared with an array-based simulator from Atos. The results showed that the decision diagram-based solution outperforms the industry-grade solutions for many quantum algorithms (e.g., for Shor's algorithm to factor integers), while using less hardware resources. Overall, this review and evaluation of quantum decision diagrams confirmed their usefulness for quantum circuit simulation, and important insights into their behavior were gained for the remainder of this book.

© The Author(s), under exclusive license to Springer Nature Switzerland AG 2025
T. Grurl et al., *Noise-Aware Quantum Circuit Simulation with Decision Diagrams*,
Synthesis Lectures on Engineering, Science, and Technology,
https://doi.org/10.1007/978-3-031-71036-0_8

Based on these insights, an optimized decision diagram-based representation of density matrices was proposed next. Density matrices are essential when working with noisy quantum states, as they can accurately represent noisy quantum states. Although established quantum decision diagram schemes can directly represent density matrices, the Hermitian property of density matrices had not yet been exploited. To change this, a dedicated decision diagram structure was proposed to represent density matrices, which exploits this property. An empirical evaluation of the optimized decision diagram structure confirmed its advantage compared to the established decision diagram representation of density matrices. The evaluation showed that the optimized representation is up to 50% more compact, and the smaller size results in runtime improvements for quantum circuit simulation of up to 53%.

Building on the established foundation, two complementary approaches for decision diagram-based noise-aware quantum circuit simulation were investigated next. First, a deterministic approach was considered, which uses density matrices and Kraus operators to model noise effects during quantum circuit simulation. The effects of this formalism for decision diagram-based quantum circuit simulation were evaluated, and an advanced scheme was proposed to realize the necessary functionality in an efficient manner. The proposed solution was then implemented in C++ and thoroughly evaluated. The results confirmed the usefulness of the proposed optimizations, which improved the performance considerably compared to the straightforward implementation for all the considered quantum circuits. However, when comparing the proposed solution with industry-grade simulators, it was only considerably faster for one of the considered quantum applications, while it was slower for most others.

To address the shortcoming of the deterministic approach, the next considered approach avoided the use of density matrices and instead considered errors in a stochastic fashion. In doing so, the deterministic description of the quantum state was lost, and the quantum state was instead approximated by forming empirical averages over multiple simulation runs (Monte Carlo). An advanced decision diagram-based stochastic quantum circuit simulator was proposed, implemented in C++, and extensively evaluated. The evaluation not only confirmed the usefulness of the proposed optimizations, but also showed tremendous improvements (speed-ups of several magnitudes and much better scalability) compared to the state-of-the-art solutions from Atos and Qiskit for many circuits.

Finally, to illustrate the usefulness of the developed simulators, the necessity of noise-aware quantum circuit simulation was illustrated for one use-case: quantum error correction. Error-correcting codes are an essential part of building large-scale and reliable quantum hardware, and accordingly, a lot of research in this area is currently being conducted. However, currently most of the corresponding work in this domain is heavily dependent on manual labor and/or is based on theoretical results only. To support researchers and engineers working in this area, a framework was proposed that automates the evaluation of these error correction schemes. More precisely, the framework allows one to automatically apply an error correction scheme to a given quantum circuit, followed by a *noise-aware* quantum circuit simulation to evaluate its usefulness. The advantages and usefulness of the proposed framework were demonstrated by evaluating the reliability of the Steane error-correcting code with respect to different parameters.

Overall, this book confirms the utility of decision diagrams for noise-aware quantum circuit simulation, as well as the need to consider these noise effects during the development and evaluation of quantum algorithms. All tools developed within the scope of this book are available under MIT License as part of the Munich Quantum Toolkit (MQT [115]; formerly known as JKQ [114]) at https://github.com/cda-tum/.

References

1. Aaronson, Scott, and Daniel Gottesman. 2004. Improved Simulation of Stabilizer Circuits. *Physical Review A.* 70 (5): 052328. https://doi.org/10.1103/PhysRevA.70.052328.
2. Afshin Abdollahi and Massoud Pedram. 2006. Analysis and Synthesis of Quantum Circuits by Using Quantum Decision Diagrams. *Design, automation and test in Europe*, 317–322. https://doi.org/10.1109/DATE.2006.244176.
3. Acharya, Rajeev, Igor Aleiner, Richard Allen, et al. 2023. Suppressing Quantum Errors by Scaling a Surface Code Logical Qubit. *Nature* 614 (7949): 676–681. https://doi.org/10.1038/s41586-022-05434-1.
4. Aharonov, Dorit and Michael Ben-Or 1997. Fault-Tolerant Quantum Computation with Constant Error. In *Symposium on Theory of Computing*, 176–188. https://doi.org/10.1145/258533.258579.
5. Arute, Frank, Kunal Arya, Ryan Babbush, et al. 2019. Quantum Supremacy Using a Programmable Superconducting Processor. *Nature* 574 (7779): 505–510. https://doi.org/10.1038/s41586-019-1666-5.
6. Atos SE. 2019. Atos Announces World First in Quantum Computing. https://atos.net/en-na/north-america/atos-announces-world-first-in-quantum-computing. Accessed 13 Aug 2023.
7. Atos SE. 2016. Quantum Learning Machine. https://atos.net/en/solutions/quantum-learning-machine. Accessed 13 Aug 2023.
8. Bacon, Dave. 2006. Operator quantum error-correcting subsystems for self-correcting quantum memories. *Physical Review A.* 73 (1): 012340. https://doi.org/10.1103/PhysRevA.73.012340.
9. Bahar, Iris R., Erica A. Frohm, Charles M. Gaona, Gary D. Hachtel, Enrico Macii, Abelardo Pardo, and Fabio Somenzi. 1993. Algebraic Decision Diagrams and Their Applications. In *International Conference on CAD*, 188–191. https://doi.org/10.1023/A:1008699807402.
10. Benioff, Paul. 1980. The Computer as a Physical System: A Microscopic Quantum Mechanical Hamiltonian Model of Computers as Represented by Turing Machines. *Journal of Statistical Physics* 22 (5): 563–591. https://doi.org/10.1007/BF01011339.
11. Berent, Lucas, Lukas Burgholzer, and Robert Wille. 2023. Software Tools for Decoding Quantum Low-Density Parity Check Codes. In *Asia and South Pacific Design Automation Conference*, 709–714. https://doi.org/10.1145/3566097.3567934.

© The Editor(s) (if applicable) and The Author(s), under exclusive license to Springer Nature Switzerland AG 2025
T. Grurl et al., *Noise-Aware Quantum Circuit Simulation with Decision Diagrams*,
Synthesis Lectures on Engineering, Science, and Technology,
https://doi.org/10.1007/978-3-031-71036-0

12. Bharti, Kishor, Alba Cervera-Lierta, Thi Ha Kyaw, et al. 2022. Noisy Intermediate-scale Quantum Algorithms. *Reviews of Modern Physics* 94(1): 015004. https://doi.org/10.1103/RevModPhys.94.015004.

13. Bhattacharyya, Amitava. 1946. On a Measure of Divergence between Two Multi-nomial Populations. *Sankhyā: The Indian Journal of Statistics* 7(4): 401–406.

14. Biamonte, Jacob D., and Ville Bergholm. 2017. Tensor Networks in a Nutshell. arXiv: 1708.00006.

15. Boixo, Sergio, Sergei V. Isakov, Vadim N. Smelyanskiy, et al. 2018. Characterizing Quantum Supremacy in Near-Term Devices. *Nature Physics* 14 (6): 595–600. https://doi.org/10.1038/s41567-018-0124-x.

16. Boole, George. 1847. *The Mathematical Analysis of Logic: Being an Essay Towards a Calculus of Deductive Reasoning.* Cambridge University Press. https://doi.org/10.1017/CBO9780511701337

17. Bravyi, Sergey, and Alexei Y. Kitaev. 1998. Quantum codes on a lattice with boundary. arXiv: quant-ph/9811052.

18. Bryant, Randal E. 1992. Symbolic Boolean Manipulation with Ordered Binary- Decision Diagrams. *ACM Computing Surveys* 24 (3): 293–318. https://doi.org/10.1145/136035.136043.

19. Bryant, Randal E., and Chen Yirng-An. 1995. Verification of Arithmetic Circuits with Binary Moment Diagrams. In *Design Automation Conference*, 535–541. https://doi.org/10.1109/DAC.1995.250005.

20. Burgholzer, Lukas, and Robert Wille. 2021. Advanced Equivalence Checking for Quantum Circuits. *IEEE Transaction on CAD of Integrated Circuits and Systems* 40(9): 1810–1824. https://doi.org/10.1109/TCAD.2020.3032630.

21. Calderbank, Robert A., and Peter W. Shor. 1996. Good Quantum Error-correcting Codes Exist. *Physical Review A.* 54 (2): 1098–1105. https://doi.org/10.1103/PhysRevA.54.1098.

22. Cao, Yudong, Jonathan Romero, Jonathan P. Olson, et al. 2019. Quantum Chemistry in the Age of Quantum Computing. *Chemical Reviews* 119 (19): 10856–10915. https://doi.org/10.1021/acs.chemrev.8b00803.

23. Capdevila-Cortada, Marcal. 2019. Electrifying the Haber-Bosch. *Nature Catalysis* 2 (12): 1055. https://doi.org/10.1038/s41929-019-0414-4.

24. Cheng, Song, Chenfeng Cao, Chao Zhang, Yongxiang Liu, Shi-Yao. Hou, Xu. Pengxiang, and Bei Zeng. 2021. Simulating Noisy Quantum Circuits with Matrix Product Density Operators. *Physical Review Research* 3 (2): 023005. https://doi.org/10.1103/PhysRevResearch.3.023005.

25. Cirq Developers. 2022. Cirq. Version v1.2.0. In *Zenodo.* https://doi.org/10.5281/zenodo.8161252.

26. Cross, Andrew, Ali Javadi-Abhari, Thomas Alexander, et al. 2022. OpenQASM 3: A Broader and Deeper Quantum Assembly Language. *ACM Transactions on Quantum Computing* 3 (3): 1–50. https://doi.org/10.1145/3505636.

27. Raedt, De., Fengping Jin Hans, Dennis Willsch, Madita Willsch, Naoki Yoshioka, Nobuyasu Ito, Shengjun Yuan, and Kristel Michielsen. 2019. Massively Parallel Quantum Computer Simulator, Eleven Years Later. *Computer Physics Communications* 237: 47–61. https://doi.org/10.1016/j.cpc.2018.11.005.

28. Dennis, Eric, Alexei Kitaev, Andrew Landahl, and John Preskill. 2002. Topological Quantum Memory. *Journal of Mathematical Physics* 43 (9): 4452–4505. https://doi.org/10.1063/1.1499754.

29. Deutsch, David, and Richard Jozsa. 1907. Rapid Solution of Problems by Quantum Computation. *Royal Society of London Series A.* 1992: 553–558. https://doi.org/10.1098/rspa.1992.0167.

30. Devoret, Michael, and Robert Schoelkopf. 2013. Superconducting Circuits for Quantum Information: An Outlook. *Science* 339: 1169–1174. https://doi.org/10.1126/science.1231930.

31. Drechsler, Rolf, Andisheh Sarabi, Michael Theobald, Bruno Becker, and Marek A. Perkowski. 1994. Efficient Representation and Manipulation of Switching Functions Based on Ordered Kronecker Functional Decision Diagrams. In *Design Automation Conference*, 415–419. https://doi.org/10.1145/196244.196444.

32. Eastin, Bryan, and Emanuel Knill. 2009. Restrictions on Transversal Encoded Quantum Gate Sets. *Physical review letters* 102: 110502. https://doi.org/10.1103/PhysRevLett.102.110502.

33. Farhi, Edward, Jeffrey Goldstone, Sam Gutmann, and Leo Zhou. 2022. The Quantum Approximate Optimization Algorithm and the Sherrington-Kirkpatrick Model at Infinite Size. *Quantum* 6: 759. https://doi.org/10.22331/q-2022-07-07-759.

34. Feinstein, David Y., and Mitchell A. Thornton. 2012. Reversible Logic Synthesis Based on Decision Diagram Variable Ordering. In *Multiple-Valued Logic and Soft Computing*, 325–339.

35. Feinstein, David Y., Mitchell A. Thornton, and D. Michael Miller. 2008. On the Data Structure Metrics of Quantum Multiple-Valued Decision Diagrams. In *International Symposium on Multi-Valued Logic*, 138–143. https://doi.org/10.1109/ISMVL.2008.28.

36. Felter, Wes, Alexandre Ferreira, Ram Rajamony, and Juan Rubio. 2015. An Updated Performance Comparison of Virtual Machines and Linux Containers. In *International Symposium on Performance Analysis of Systems and Software*, 171–172. https://doi.org/10.1109/ISPASS.2015.7095802.

37. Feynman, Richard P. 1982. Simulating physics with computers. *International Journal of Theoretical Physics*, 467–488. https://doi.org/10.1007/BF02650179.

38. Fletcher, Andrew S., Peter W. Shor, and Moe Z. Win. 2008. Structured Near-optimal Channel-adapted Quantum Error Correction. *Physical Review A.* 77 (1): 012320. https://doi.org/10.1103/PhysRevA.77.012320.

39. Fowler, Austin G., Ashley M. Stephens, and Peter Groszkowski. 2009. High-Threshold Universal Quantum Computation on the Surface Code. *Physical Review A.* 80 (5): 052312. https://doi.org/10.1103/PhysRevA.80.052312.

40. Gergov, Jordan, and Christoph Meinel. 1994. Efficient Boolean Manipulation with OBDD's can be Extended to FBDD's. *IEEE Transactions on Computers* 43 (10): 1197–1209. https://doi.org/10.1109/12.324545.

41. Gheorghiu, Vlad. 2018. Quantum++: A Modern C++ Quantum Computing Library. *PLOS ONE* 13 (12): 1–27. https://doi.org/10.1371/journal.pone.0208073.

42. Gidney, Craig. 2021. Stim: A fast stabilizer circuit simulator. *Quantum* 5: 497. https://doi.org/10.22331/q-2021-07-06-497.

43. Gilchrist, Alexei, Nathan K. Langford, and Michael A. Nielsen. 2005. Distance Measures to Compare Real and Ideal Quantum Processes. *Physical Review A.* 71 (6): 062310. https://doi.org/10.1103/PhysRevA.71.062310.

44. Goodmann, David, Thornton A. Mitchell, David Y. Feinstein, and D. Michael Miller. 2007. Quantum Logic Circuit Simulation Based on the QMDD Data Structure. In *International Reed-Muller Workshop*, 99–105.

45. Gottesman, Daniel. 1996. Class of Quantum Error-correcting Codes Saturating the Quantum Hamming Bound. *Physical Review A.* 54 (3): 1862–1868. https://doi.org/10.1103/PhysRevA.54.1862.

46. Grover, Lov K. 1996. A Fast Quantum Mechanical Algorithm for Database Search. In *Symposium on Theory of Computing*, 212–219. https://doi.org/10.1145/237814.237866.

47. Grurl, Thomas, Jürgen Fuß, Stefan Hillmich, Lukas Burgholzer, and Robert Wille. 2020. Arrays vs. Decision Diagrams: A Case Study on Quantum Circuit Simulators. In *International Symposium on Multi-Valued Logic*, 176–181. https://doi.org/10.1109/ISMVL49045.2020.000-9.

48. Grurl, Thomas, Jürgen Fuß, and Robert Wille. 2020. Considering Decoherence Errors in the Simulation of Quantum Circuits Using Decision Diagrams. In *International Conference on CAD*, 1–7. https://doi.org/10.1145/3400302.3415622.

49. Grurl, Thomas, Jürgen Fuß, and Robert Wille. Lessons Learnt in the Implementation of Quantum Circuit Simulation Using Decision Diagrams. In *International Symposium on Multi-Valued Logic*, 87–92. https://doi.org/10.1109/ISMVL51352.2021.00024.

50. Grurl, Thomas Jürgen Fuß, and Robert Wille. 2022. Noise-aware Quantum Circuit Simulation With Decision Diagrams. *IEEE Transaction on CAD of Integrated Circuits and Systems*, 860–873. https://doi.org/10.1109/TCAD.2022.3182628.

51. Grurl, Thomas, Jürgen Fuß, and Robert Wille. 2023. Optimized Density Matrix Representations: Improving the Basis for Noise-Aware Quantum Circuit Design Tools. In *International Symposium on Multi-Valued Logic*, 141–146. https://doi.org/10.1109/ISMVL57333.2023.00036.

52. Grurl, Thomas, Richard Kueng, Jürgen Fuß, and Robert Wille. 2021. Stochastic Quantum Circuit Simulation Using Decision Diagrams. In *Design, Automation and Test in Europe*, 194–199. https://doi.org/10.23919/DATE51398.2021.9474135.

53. Grurl, Thomas, Christoph Pichler, Jürgen Fuß, and Robert Wille. 2023. Automatic Implementation and Evaluation of Error-Correcting Codes for Quantum Computing: An Open-Source Framework for Quantum Error-Correction. In *VLSI Design*, 301–306. https://doi.org/10.1109/VLSID57277.2023.00068.

54. Giacomo Guerreschi, Gian, Justin Hogaboam, Fabio Baruffa, and Nicolas P. D. Sawaya. 2020. Intel Quantum Simulator: A Cloud-ready High-performance Simulator of Quantum Circuits. *Quantum Science and Technology* 5: 034007. https://doi.org/10.1088/2058-9565/ab8505.

55. Hillmich, Stefan, Lukas Burgholzer, Florian Stögmüller, and Robert Wille. 2022. Reordering Decision Diagrams for Quantum Computing Is Harder Than You Might Think. In *International Conference of Reversible Computation*, 93–107. https://doi.org/10.1007/978-3-031-09005-9_7.

56. Hillmich, Stefan, Richard Kueng, Igor L. Markov, and Robert Wille. 2021. As Accurate as Needed, as Efficient as Possible: Approximations in DD-based Quantum Circuit Simulation. In *Design, Automation and Test in Europe*. 188–193. https://doi.org/10.23919/DATE51398.2021.9474034.

57. Hillmich, Stefan, Alwin Zulehner, and Robert Wille. 2020. Concurrency in DD-based Quantum Circuit Simulation. In *Asia and South Pacific Design Automation Conference*, 115–120. https://doi.org/10.1109/ASP-DAC47756.2020.9045711.

58. Hoeffding, Wassily. 1963. Probability Inequalities for Sums of Bounded Random Variables. *Journal of the American Statistical Association* 58 (301): 13–30. https://doi.org/10.1080/01621459.1963.10500830.

59. Hsieh, Min-Hsiu., and François Le Gall. 2011. NP-Hardness of Decoding Quantum Error-correction Codes. *Physical Review A*. 83 (5): 052331. https://doi.org/10.1103/PhysRevA.83.052331.

60. Huang, Hsin-Yuan., Richard Kueng, and John Preskill. 2021. Information-Theoretic Bounds on Quantum Advantage in Machine Learning. *Physical Review Letters* 126 (19): 190505. https://doi.org/10.1103/PhysRevLett.126.190505.

61. IBM Newsroom. 2022. IBM Unveils 400 Qubit-Plus Quantum Processor and Next-Generation IBM Quantum System Two. https://newsroom.ibm.com/2022-11-09-IBM-Unveils-400-Qubit-Plus-Quantum-Processor-and-Next-Generation-IBM-Quantum-System-Two. Accessed 20 Feb 2023.

62. IBM Unveils World's First Integrated Quantum Computing System for Commercial Use. https://newsroom.ibm.com/2019-01-08-IBM-Unveils-Worlds-First-Integrated-Quantum-Computing-System-for-Commercial-Use. Accessed 05 Apr 2023.

63. IBM Quantum. 2021. https://quantum-computing.ibm.com/.

64. Jones, Tyson, Anna Brown, Ian Bush, and Simon C Benjamin. 2019. QuEST and High Performance Simulation of Quantum Computers. *Scientific Reports* 9(1): 1–11. https://doi.org/10.1038/s41598-019-47174-9.

65. Jozsa, Richard. 1994. Fidelity for Mixed Quantum States. *Journal of Modern Optics* 41 (12): 2315–2323. https://doi.org/10.1080/09500349414552171.

66. Nader Khammassi, Imran Ashraf, Xiang Fu, Carmen Almudever, and Koen Bertels. 2017. QX: A High-Performance Quantum Computer Simulation Platform. In *Design, Automation and Test in Europe*, 464–469. https://doi.org/10.23919/DATE.2017.7927034.

67. Kitaev, Alexei Y. 1997. Quantum Computations: Algorithms and Error Correction. *Russian Mathematical Surveys* 52 (6): 1191–1249. https://doi.org/10.1070/RM1997v052n06ABEH002155.

68. Knill, Emanuel, and Raymond Laflamme. 1997. Theory of Quantum Error-correcting Codes. *Physical Review A.* 55 (2): 900–911. https://doi.org/10.1103/PhysRevA.55.900.

69. Knill, Emanuel, Raymond Laflamme, and Wojciech H. Zurek. 1998. Resilient Quantum Computation. *Science* 279 (5349): 342–345. https://doi.org/10.1126/science.279.5349.342.

70. Laflamme, Raymond Cesar Miquel, Juan Pablo Paz, and Wojciech Hubert Zurek. 1996. Perfect quantum error correcting code. *Physical Review Letters* 77(1): 198–201. https://doi.org/10.1103/PhysRevLett.77.198.

71. Li, Ang, Samuel Stein, Sriram Krishnamoorthy, and James Ang. 2023. QASMBench: A Low-Level Quantum Benchmark Suite for NISQ Evaluation and Simulation. *ACM Transactions on Quantum Computing* 4 (2): 1–26. https://doi.org/10.1145/3550488.

72. Lieuwe, Vinkhuijzen, Grurl Thomas, Hillmich Stefan, Brand Sebastiaan, Robert Wille, and Laarman Alfons. 2023. Efficient Implementation of LIMDDs for Quantum Circuit Simulation. In *International Symposium on Model Checking of Software*.

73. Mandrà, Salvatore , Jeffrey Marshall, Eleanor G. Rieffel, and Rupak Biswas. 2021. HybridQ: A Hybrid Simulator for Quantum Circuits. In *International Workshop on Quantum Computing Software*, 99–109. https://doi.org/10.1109/QCS54837.2021.00015.

74. Markov, Igor L., and Yaoyun Shi. 2008. Simulating Quantum Computation by Contracting Tensor Networks. *SIAM Journal on Computing* 38 (3): 963–981. https://doi.org/10.1137/050644756.

75. Merkel, Dirk. 2014. Docker: Lightweight Linux Containers for Consistent Development and Deployment. *Linux Journal* 2014(239).

76. Miller, D. Michael, and Rolf Drechsler. (2003). Augmented Sifting Of Multiple-valued Decision Diagrams. In *International Symposium on Multi-Valued Logic*, 375–382.https://doi.org/10.1109/ISMVL.2003.1201431.

77. Miller, D. Michael, David D. Feinstein, and Mitchell A. Thornton. 2007. QMDD Minimization using Sifting for Variable Reordering. *Multiple-Valued Logic and Soft Computing* 13: 1–16.

78. Miller, D. Michael, David Y. Feinstein, and Mitchell A. Thornton. 2007. Variable Reordering and Sifting for QMDD. In *International Symposium on Multi-Valued Logic*, 10–10. https://doi.org/10.1109/ISMVL.2007.59.

79. Miller, D. Michael, Thornton A. Mitchell, and David Goodmann. 2006. A Decision Diagram Package for Reversible and Quantum Circuit Simulation. In *International Conference on Evolutionary Computation*, 8597–8604. https://doi.org/10.1109/CEC.2006.1688610.

80. Miller, D. Michael, and Mitchell A. Thornton. 2006. QMDD: A Decision Diagram Structure for Reversible and Quantum Circuits. In *International Symposium on Multi-Valued Logic*. https://doi.org/10.1109/ISMVL.2006.35.

81. Minato, Shin-ichi. 1993. Zero-Suppressed BDDs for Set Manipulation in Combinatorial Problems. In *Design Automation Conference Association for Computing Machinery*, 272–277. https://doi.org/10.1145/157485.164890.

82. Nielsen, Michael, and Issac Chuang. 2010. *Quantum Computation and Quantum Information*. Cambridge University Press. https://doi.org/10.1017/CBO9780511976667.

83. Niemann, Philipp, Robert Wille, and Rolf Drechsler. 2018. Efficient Synthesis of Quantum Circuits Implementing Clifford Group Operations. In *Asia and South Pacific Design Automation Conference*, 483–488. https://doi.org/10.1109/ASPDAC.2014.6742938.

84. Niemann, Philipp, Robert Wille, and Rolf Drechsler. 2014. Equivalence Checking in Multi-level Quantum Systems. In *International Conference of Reversible Computation*, 201–215. https://doi.org/10.1007/978-3-319-08494-7_16.

85. Niemann, Philipp, Robert Wille, D. Michael Miller, Mitchell A. Thornton, and Rolf Drechsler. 2016. QMDDs: Efficient Quantum Function Representation and Manipulation. *IEEE Transaction on CAD of Integrated Circuits and Systems* 35(1): 86–99. https://doi.org/10.1109/TCAD.2015.2459034.

86. Orús, Román. 2014. A Practical Introduction to Tensor Networks: Matrix Product States and Projected Entangled Pair States. *Annals of Physics* 349: 117–158. https://doi.org/10.1016/j.aop.2014.06.013.

87. Park, Daeyoung Heehoon Kim, Jinpyo Kim, Taehyun Kim, and Jaejin Lee. 2022. SnuQS: Scaling Quantum Circuit Simulation Using Storage Devices. In *International Conference on Supercomputing*, 1–13. https://doi.org/10.1145/3524059.3532375.

88. Preskill, John. 2018. Quantum Computing in the NISQ Era and Beyond. *Quantum* 2: 79. https://doi.org/10.22331/q-2018-08-06-79.

89. Qiskit contributors. 2023. Qiskit: An Open-source Framework for Quantum Computing. https://doi.org/10.5281/zenodo.2573505.

90. Nils, Quetschlich, Lukas Burgholzer, and Robert Wille. 2023. MQT Bench: benchmarking Software and Design Automation Tools for Quantum Computing. In *Quantum*, p 1062. https://doi.org/10.22331/q-2023-07-20-1062.

91. Ristè, Diego, Marcus P. da Silva, Colm A. Ryan, Andrew W. Cross, Antonio D. Córcoles, John A. Smolin, Jay M. Gambetta, Jerry M. Chow, and Blake R. Johnson. 2017. Demonstration of Quantum Advantage in Machine Learning. *npj Quantum Information* 3(1): 16. https://doi.org/10.1038/s41534-017-0017-3.

92. Robertson, Alan, Christopher Granade, Stephen D. Bartlett, and Steven T. Flam-mia. 2017. Tailored Codes for Small Quantum Memories. *Physical Review A.* 8 (6): 064004. https://doi.org/10.1103/PhysRevApplied.8.064004.

93. Roffe, Joschka. 2019. Quantum Error Correction: An Introductory Guide. *Contemporary Physics* 60 (3): 226–245. https://doi.org/10.1080/00107514.2019.1667078.

94. Samoladas, Vasilis. Improved BDD Algorithms for the Simulation of Quantum Circuits. In *European Symposium on Algorithms*, 720–731. https://doi.org/10.1007/978-3-540-87744-8_60.

95. Shor, Peter W. 1994. Algorithms for Quantum Computation: Discrete Logarithms and Factoring. In *Symposium on Foundations of Computer Science*, 124–134. https://doi.org/10.1109/SFCS.1994.365700.

96. Shor, Peter W. 1997. Polynomial-Time Algorithms for Prime Factorization and Discrete Logarithms on a Quantum Computer. *SIAM Journal of Computer* 26 (5): 1484–1509. https://doi.org/10.1137/S0097539795293172.

97. Shor, Peter W. 1995. Scheme for Reducing Decoherence in Quantum Computer Memory. *Physical Review A.* 52 (4): R2493–R2496. https://doi.org/10.1103/PhysRevA.52.R2493.

98. Smelyanskiy, Mikhail, Nicolas P. D. Sawaya, and Alán Aspuru-Guzik. 2016. qHiPSTER: The Quantum High Performance Software Testing Environment. arXiv:1601.07195.

99. Smith, Kaitlin N., and Mitchell A. Thornton. 2019. A Quantum Computational Compiler and Design Tool for Technology-Specific Targets. In *International Symposium on Computer Architecture*, 579–588. https://doi.org/10.1145/3307650.3322262.

100. Smith, Kaitlin N., and Mitchell A. Thornton. 2019. Quantum Logic Synthesis with Formal Verification. In *Midwest Symposium on Circuits and Systems*, 73–76. https://doi.org/10.1109/MWSCAS.2019.8885132.

101. Steane, M. Andrew. 1996. Error Correcting Codes in Quantum Theory. *Physical Review Letters* 77(5): 793–797. https://doi.org/10.1103/PhysRevLett.77.793.

102. Steffen, Matthias, Jerry Chow, Sarah Sheldon, and Doug McClure. 2022. A New Eagle in the Poughkeepsie Quantum Datacenter: IBM Quantum's Most Performant System Yet. https://research.ibm.com/blog/eagle-quantum-error-mitigation. Accessed 08 Aug 2023.

103. Steiger, Damian, Thomas Hüner, and Matthias Troyer. 2018. ProjectQ: An Open Source Software Framework for Quantum Computing. *Quantum* 2. https://doi.org/10.22331/q-2018-01-31-49.

104. Tannu, Swamit S., and Moinuddin K. Qureshi. Not All Qubits Are Created Equal: A Case for Variability-Aware Policies for NISQ-Era Quantum Computers. In *International Conference on Architectural Support for Programming Languages and Operating Systems*, 987–999. https://doi.org/10.1145/3297858.3304007.

105. Viamontes, George F., Igor L. Markov, and John P. Hayes. 2004. Graph-Based Simulation of Quantum Computation in the Density Matrix Representation. *Quantum Information Computing* 5 (2): 113–130. https://doi.org/10.1117/12.542767.

106. Viamontes, George F., Igor L. Markov, and John P. Hayes. 2004. High-performance QuIDD-based Simulation of Quantum Circuits. In *Design, Automation and Test in Europe*, 1354–1355. https://doi.org/10.1109/DATE.2004.1269084.

107. Viamontes, George F., Igor L. Markov, and John P. Hayes. 2009. *Quantum Circuit Simulation*. Springer. https://doi.org/10.1007/978-90-481-3065-8.

108. Vidal, Guifré. 2003. Efficient Classical Simulation of Slightly Entangled Quantum Computations. *Physical Review Letters* 91 (14): 147902. https://doi.org/10.1103/PhysRevLett.91.147902.

109. Villalonga, Benjamin, Sergio Boixo, Bron Nelson, et al. 2019. A Flexible High-performance Simulator For Verifying and Benchmarking Quantum Circuits Implemented on Real Hardware. *npj Quantum Information* 5(1). https://doi.org/10.1038/s41534-019-0196-1.

110. Wang, Shiou-An, Chin-Yung Lu, I-Ming Tsai, and Sy-Yen Kuo. 2008. An XQDD-Based Verification Method for Quantum Circuits. *IEICE Transactions on Fundamentals of Electronics, Communications and Computer Sciences* E91-A(2): 584–594. https://doi.org/10.1093/ietfec/e91-a.2.584.

111. Wecker, Dave, and Krysta Svore, 2024. LIQUi: A Software Design Architecture and Domain-Specific Language for Quantum Computing. arXiv: 1402.4467.

112. Wegener, Ingo. 2000. Branching Programs and Binary Decision Diagrams. *Society for Industrial and Applied Mathematics*. https://doi.org/10.1137/1.9780898719789.

113. Wille, Robert, Lukas Burgholzer, Stefan Hillmich, Thomas Grurl, Alexander Ploier, and Thomas Peham. 2022. The Basis of Design Tools for Quantum Computing: Arrays, Decision Diagrams, Tensor Networks, and ZX-Calculus. In: *Design Automation Conference* 1367–1370. https://doi.org/10.1145/3489517.3530627.

114. Wille, Robert, Stefan Hillmich, and Lukas Burgholzer. 2020. JKQ: JKU Tools for Quantum Computing. In *International Conference on CAD*.

115. Wille, Robert, Stefan Hillmich, and Lukas Burgholzer. 2022. MQT: The Munich Quantum Toolkit. In *Gesellschaft für Informatik Quantum Computing Workshop*

116. Wootters, William K., and Wojciech H. Zurek. 1982. A Single Quantum Cannot be Cloned. *Nature* 299 (5886): 802–803. https://doi.org/10.1038/299802a0.

117. Wootton, James R., Andreas Peter, János. R. Winkler, and Daniel Loss. 2017. Proposal for a Minimal Surface Code Experiment. *Physical Review A.* 96 (3): 032338. https://doi.org/10.1103/PhysRevA.96.032338.

118. Zulehner, Alwin, Stefan Hillmich, Igor Markov, and Robert Wille. 2020. Approximation of Quantum States Using Decision Diagrams. In *Asia and South Pacific Design Automation Conference*, 121–126. https://doi.org/10.1109/ASP-DAC47756.2020.9045454.

119. Zulehner, Alwin, Stefan Hillmich, and Robert Wille. 2019. How to Efficiently Handle Complex Values? Implementing Decision Diagrams for Quantum Computing. In: *International Conference on CAD*, 1–7. https://doi.org/10.1109/ICCAD45719.2019.8942057.

120. Zulehner, Alwin, Philipp Niemann, Rolf Drechsler, and Robert Wille. 2019. Accuracy and Compactness in Decision Diagrams for Quantum Computation. In: *Design, Automation and Test in Europe*, 280–283. https://doi.org/10.23919/DATE.2019.8715040.

121. Zulehner, Alwin, and Robert Wille. 2018. Advanced Simulation of Quantum Computations. *IEEE Transaction on CAD of Integrated Circuits and Systems* 38(5): 848–859. https://doi.org/10.1109/TCAD.2018.2834427.

122. Zulehner, Alwin, and Robert Wille. 2019. Matrix-Vector vs. Matrix-Matrix Multiplication: Potential in DD-based Simulation of Quantum Computations. Design, Automation and Test in Europe, 90–95. https://doi.org/10.23919/DATE.2019.8714836.

123. Zulehner, Alwin, and Robert Wille. 2018. One-Pass Design of Reversible Circuits: Combining Embedding and Synthesis For Reversible Logic. *IEEE Transaction on CAD of Integrated Circuits and Systems* 37(5): 996–1008. https://doi.org/10.1109/TCAD.2017.2729468.